ROYAL NAVAL CORDITE FACTORY
HOLTON HEATH

ROYAL NAVAL CORDITE FACTORY
HOLTON HEATH

Malcolm Bowditch
& Leslie Hayward

FOLLY BOOKS

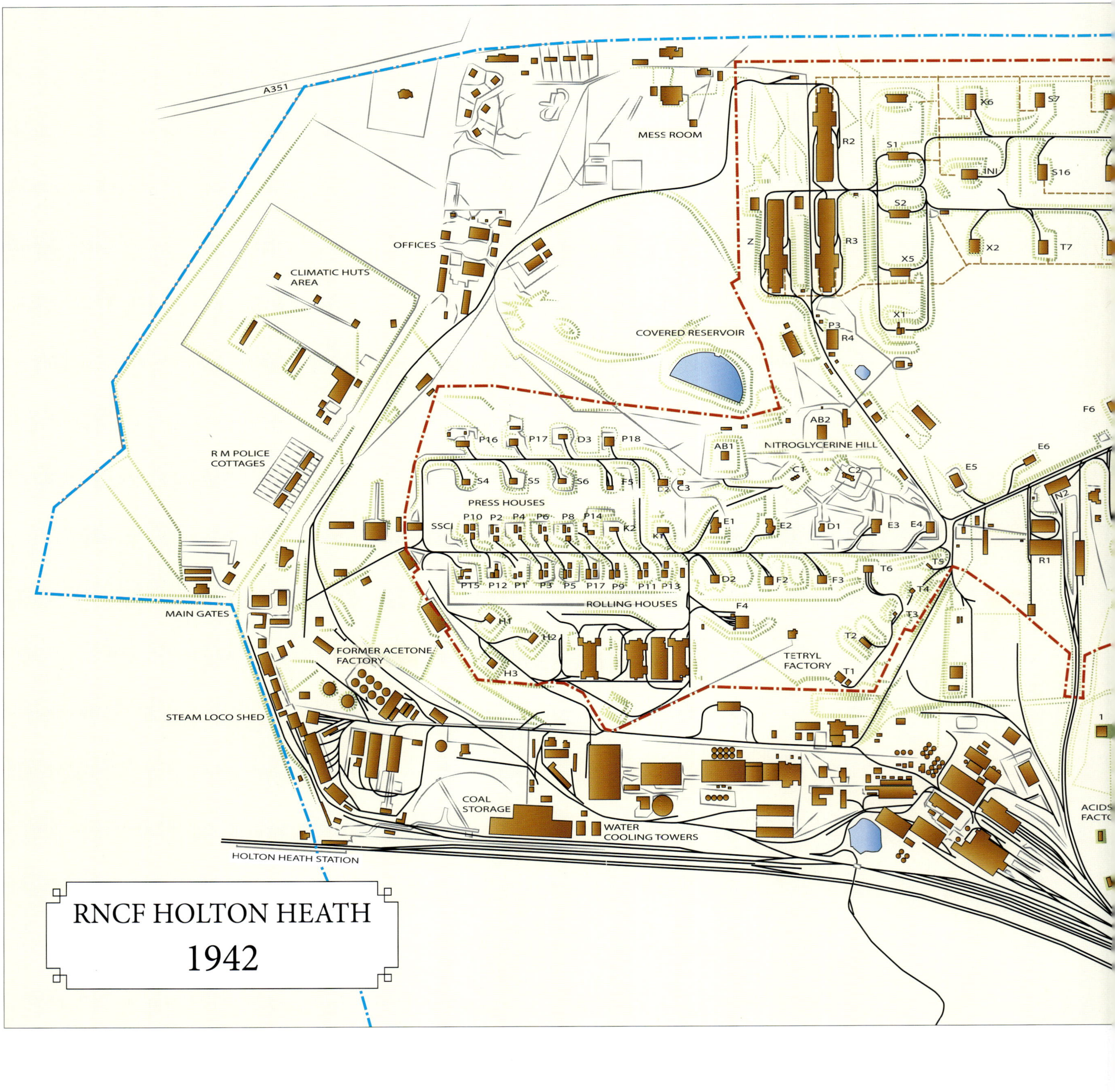

RNCF HOLTON HEATH
1942

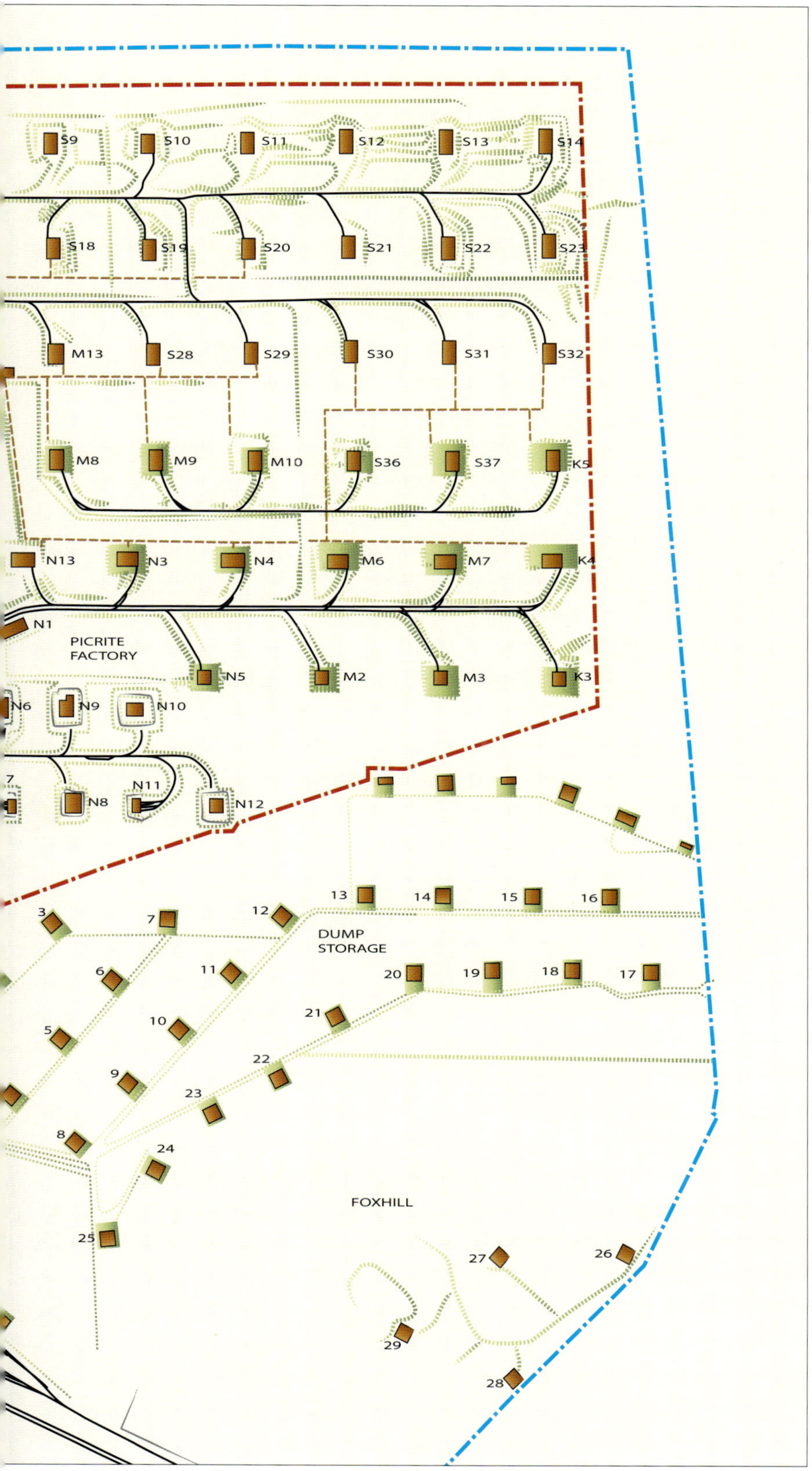

First published by Finial Publishing and Printing Services 1996

Folly Books 2015

A catalogue record for this book is available from the British Library.

ISBN 978-0-9928554-4-4

Published by Folly Books Ltd
Monkton Farleigh
BA15 2QP
www.follybooks.co.uk

Designed and typeset by Vicky

Printed and bound in India by Replika Press Pvt Ltd

Introduction to the Second Edition

The first edition of this book was well received and completely sold out. The authors had been approached on a number of occasions to consider a reprint, but from the initial date of publication a number of things had changed which resulted in some of the original writing, although factual at the time, requiring some modification to record changed circumstances.

A further consideration was the accumulation over time of additional photographic material and in particular, the discovery of a large album containing some two hundred and forty photographs, most of which were quite new to us. Finally, as must often be the case, once a book is published, additional material becomes available, too late for our inclusion in the first edition, but which can be used now.

This edition then, whilst retaining the original work, has been substantially updated with developments which have a bearing on our subject, yet which occurred after the original date of publication.

Engineering terms throughout this account have been kept to those in common use at the time, i.e. Imperial.

Malcolm Bowditch
& Leslie Hayward

Acknowledgements

The authors wish to record their thanks to the following people who have assisted in the writing of this account:

Mr C W Alexander, Mr W Austin, Mr F Ayres, Mr R Bailey, Mr Percy Bartlett, Mr K R Brown, Mr R L Brown, Mr D Burgess, Mr J A Burgess, Mr Ted Cake Mr Hamish Christie, Mr W H Coombes, Mr R Cox, Dr Bob Dukes, Mr N A Ford, Mr Eddie Grist, Mr W Groves, Mr F Jones, Miss P Haslar, Mr R Hillman, Dr R J Huggins, Dr Lee-Smith, Mr M McLaren, Mr C F McCourt, Mr Medwyn Parry, Mr Alan Penny, Mr T Morgan, Mr R A Panting, Mr Fred Ryder, Mr C Selby, Mr M Singleton, Mr Spear, Dr H G Stubbings, Mr W Terrell, Dr D A Tod, Mr Mike Tuck, Dr N J Wadsworth, Mr J Wallace, Mr & Mrs J B Walters, Dr E R Ward and Mr J Way.

A special thanks to Mr J V B Gray who has acted virtually as technical consultant and whose excellent memory, detailed knowledge and good humour has proved invaluable.

We also wish to acknowledge the help given by these organisations:

Admiralty Research Establishment (later DERA, Holton Heath), BAE Systems, British Land Ltd, The Ministry of Defence, The Nature Conservancy Council, the Naval Ordnance Museum, Priddy's Hard, the Royal Ordnance Factory plc, Royal Air Force and Wessex Water Authority.

Thanks go to Nick and Vicky McCamley for taking on the task of publishing this edition.

Finally, our thanks go to the volunteers of the RNCF association whose kind efforts assisted in making the first edition such a success.

CONTENTS

Chapter 1

THE BEGINNINGS

It was during the first few weeks of the First World War that Mr Winston Churchill, then First Lord of the Admiralty, insisted on the need for the Royal Navy to have its own independent supply of propellant (cordite) for shells. It was realised that propellant manufactured for use in fighting at sea had to be of a very high quality with a greater uniformity of ballistics than was necessary in propellants made for use on land. The production of such high grade propellant could best be achieved, it was decided, by a specialist factory designed specifically for this purpose.

Cordite is essentially a mixture of two explosive and potentially very hazardous materials: nitroglycerine, a dense colourless liquid, and guncotton, a fibrous solid. The large scale production of explosive calls for the very careful selection of a suitable manufacturing site and it was with this in mind that a committee from London visited Holton Heath in east Dorset during 1914. Included in this party, which doubtless inspected a number of other candidate sites, was Mr W T Thomson who was later to become the first Manager and Deputy Superintendent (MDS) of the new installation which was to be called the Royal Naval Cordite Factory (RNCF). It is probable that Colonel Sir F L Nathan, KBE, was also amongst the group that visited the site on that occasion as he had been asked by the Admiralty to take over the design and erection of the new factory.

Frederick Lewis Nathan was a leading authority in the chemical engineering and technology associated with high explosives at the time with an association with cordite which began in the pioneering days of the mid-1880s. In 1889 he joined the Royal Gunpowder Factory at Waltham Abbey and was closely associated with the earliest production of the first cordite, Cordite Mark I, in 1890. It was a happy coincidence for this country that in 1914 Sir Frederick had just about completed the design and erection of a new TNT and Propellant Factory for the Nobel's Explosives Company, for whom he had been working since he left the government service in 1909. This meant that he was able to make available a complete set of up-to-date plans for a modern factory of the type that was required.

Exactly what advantages the Holton Heath site possessed over its competitors we do not know but certainly it had a number of features that were obviously attractive. It was, for example, a reasonably level site and it did possess, at the approximate centre, hills which would be suitable for a reservoir and also for the nitroglycerine plant. In those days it was the usual practice to build nitroglycerine factories on a slope in order that the liquid product could flow gently in lead gutters from one process to another under the influence of gravity. An arrangement of this kind was the simplest and safest that could be devised for handling this dangerously sensitive material.

It also made good economic sense because all the raw materials used in nitroglycerine manufacture, and all the products, were in liquid form and to allow materials to flow under the influence of gravity was cheaper and more reliable than was the dependence on pumping systems. In addition the site had many other attractions; it was, for example, adjacent to the London & South Western Railway and there was also ready access to the backwaters of Poole harbour. Road access was also convenient with the main Wareham to Poole road passing immediately to the north-west. The site, which was essentially barren acid heathland, had very little agricultural value and, most importantly, although it was sufficiently far from centres of population to be safe it was not so remote that a suitably large workforce could not be mustered.

In any event the decision was made and the land was purchased from the owner, Sir John Lees, Bart. of Lytchett Minster, and construction started without delay. Sir Douglas Fox & Partners, later to become Freeman, Fox & Company, was retained as consulting engineers whilst the main contractor for the building of the factory was the London-based civil engineering firm of Topham, Jones & Railton. Two of the more important subcontractors were Mann Egerton Ltd who was responsible for the electrical installations, and Sturtevant Ltd who handled all the pipework and plumbing inevitably associated with the very sophisticated chemical engineering which was such an important part of this high priority project. Various rusting fittings bearing the subcontractors' names are to be found on the site to this day

RNCF Management Group

1	Mr Butlin	5	Mr Roberts	9	Mr Mr Frost	13	Capt Hammond
2	Mr Bunker	6	Mr Whitaker	10	Mr Blair	14	Dr Knight
3	Mr Hawes	7	Mr Ledbury	11	Dr Thaysen	15	Mr Gallen
4	Mr Brooksbank	8	Mr Read	12	Mr Thomson	16	Mr Hazel

An interesting sidelight on industrial relations in those early wartime days is provided by a story which involved Mr Railton, the junior partner of the main contractors. Apparently one day, when walking over the site, he came across a workman who seemed to be spending most of his time leaning on a shovel. Mr Railton approached the man and asked him who he worked for. 'Topham, Jones and that other bugger,' came the reply. The management responded instantly and told the workman to go to the office and tell them that 'that other bugger' had given him the sack. Although there was a three day conference in March of 1915 during which the government and the trades unions agreed that there would be no stoppage of work upon munitions or any other work necessary for the satisfactory completion of the war, the dismissal still seems rather peremptory by today's standards.

Later in 1915 whilst construction was still under way there was a more serious and blatant transgression of this agreement and there was an actual strike. Wartime conditions allowed management to respond with a delightfully simple and direct approach to such problems. It appears that the military at Worgret Camp were contacted and shortly afterwards a detachment of soldiers marched on to the site with fixed bayonets. The dispute was thus resolved in no time at all and as far as is known no further trouble of this kind was ever experienced.

A further dramatic event occurred during the construction period when a linesman with Mann Egerton, suspected of spying, was traced firstly to a temporary generating station, situated just to the north of the main office block on a concreted area which was later to become the site of a pair of hard tennis courts, and thence to his lodgings in Poole where he was arrested. It is understood that he was taken to the Tower of London where he was later shot as a traitor.

During this construction period the site was policed by half a dozen members of the Dorset Constabulary led by a Sergeant but after completion when production started these duties were taken over by the Metropolitan Police, both men and women, who were quartered in the long wooden building which stood at Holton Heath crossroads immediately before turning left into the factory on approaching from Poole to the east. The function of the women police was mainly that of searching female employees prior to entering danger areas. Although remote from London, it was general at this time for the Metropolitan Police to protect munitions establishments; the Royal Gunpowder Factory at Waltham Abbey and the Royal Naval Armament Depot at Priddy's Hard (both very old establishments) were also under their control. This arrangement continued until the end of the First World War after which the Royal Marine Police took over.

The police were, of course, always present in that strength proportional to the task in hand and this varied over the years. In wartime the numbers employed were understandably higher and during the Second World War the Royal Marine Police were present in considerable force with a Chief Inspector controlling an organisation including a station sergeant, three sergeants and twenty-nine constables. The police had special accommodation provided in the form of terraced cottages situated on the western side of the road leading from Holton Heath crossroads to the station. These cottages were modernised over the years and were still occupied by MoD police and their families until the site was closed in the mid-1990s. The police carried .38 calibre Smith & Wesson Military and Police model revolvers.

In the earliest days of construction, wages were still being paid in gold and Bill Groves could remember when the gold was delivered at the site on the Thursday of each week. The contractor's cashier's department was then kept very busy making up workmen's wages having often already worked till the early hours of that same day calculating the payments to be made. Although we have been unable to establish just how many men were employed on the site during the course of construction the numbers must presumably have been very large, partly because of the priority placed upon its early completion and partly because of the labour intensive nature of the work in the absence of modern earth moving machinery.

In any event, it is said that on completion, as many as 500 'brickies' were discharged at one time by the contractors. Bill retained a vivid recollection of hundreds of little piles of gold on long trestle tables. The sums of money were small not only because wages were low by modern standards but also because many of the men had applied for 'subs' so frequently during the week that for some there was scarcely anything left to receive on pay day.

The complex was completed and opened in January 1916 with the acids factory being the first in production. That such an installation was designed, built and put into production in the very short timescale of not much more than a year was a remarkable achievement. In scale it must have been one of the most significant pieces of civil engineering to have been undertaken in east Dorset since the railway navvies completed that part of the permanent way passing westwards to Dorchester through Wimborne, Broadstone,

Above: One of the earliest photographs of Holton Heath, showing track laying underneath Knights Bridge. The line of this track is now taken by the road leading from the A351 into the Holton Heath industrial estate, viewed towards the direction of Poole Harbour.

Hamworthy Junction and Holton Heath in 1846.

In the early days, workers received their pay in open metal cups with coins placed in the bottom. In this way money could be checked as it was received and then the empty cups were thrown into large wicker baskets provided for the purpose. The wages and salaries of sixty years ago seem very small by today's standards but everything was also very much cheaper then and certainly workers at the RNCF were quite well paid by comparison with agricultural workers for example. In 1916, in return for working a 48 hour week a guncotton process worker received 27/- (£1.34), a cordite process worker 28/- (£1.40) and a cordite pressman 32/- (£1.60). Nitroglycerine workers received slightly more, as did those working in the acids factory. By way of lending perspective to these figures it is interesting to note that the Manager and Deputy Superintendent at about this time during the First World War would have been on a scale from £500-£650 per annum, and a senior foreman would have received £200-£300 per annum.

Initially, those workers who used trains to get to work were issued with free, quarterly railway season tickets but about half of the workforce travelled by road and they quickly pointed out that they had travelling expenses too. The season tickets were therefore abolished after a year or so and all workers were granted a 2/- (10p) per week travelling allowance. Most of the road users were, of course, cyclists at that time and for most of the life of the factory. Mrs Edna Holland well recalled the traffic hazard of encountering the egress of several hundred unlighted bicycles leaving the site during the blackout!

Apart from production workers there was a very wide range of trades practised within the factory from the beginning, not all of which are immediately obvious. There were, for example, cobblers and bootmakers, who were employed to maintain the special footwear that was supplied for workers in the danger areas, and there were laundry workers, seamstresses and searchers. There were also firemen and pattern makers and sail makers - who made and maintained protective rope mantlets that surrounded the cordite presses. In addition there were doctors and nurses and messengers and also a full-time rat catcher. The rat catcher was responsible for destroying vermin of all kinds, including squirrels and rabbits, and in the course of his duties he roamed freely over the establishment with a twelve-bore shot gun under his arm. As a consequence, he was a rich source of gossip and of rabbits, which he sold as a side-line.

The Works Department record of workers also lists tradesmen ranging from asphalters and blacksmiths through drainmen and platelayers to sawyers and roofers. There were also 'boy workers' and 'yard boys' employed extensively in the factory. In addition a number of youths were admitted annually into the RNCF in the thirties as Trade Lads. A science examination set for potential recruits in 1937 included the question 'Name three well-known gases and describe briefly the production of one of them, and the properties which distinguish it from the others'. One of a batch of these papers found in a roof space at Holton Heath contained the following answer to this question, 'Three well-known gases are, the gas which is used for light, gas which is used in hospitle for oporating, (sic) and gas which is only used in war time'. Since this was the fuller of only two answers attempted it may be reasonably assumed that this candidate failed to reach the required standard!

A clockmaker was also employed, and his job was to keep all the clocks in the factory wound up and well maintained. Apart from clocks in buildings outside the Danger Area fence there were also large clocks, of the sort commonly seen on railway stations, mounted on gantries at intervals along the railway cuttings within the Danger Area. For a period in the thirties, the clockmaker was Edward 'John' Bull who, on one occasion, was kneeling down on a path where it crossed a standard-gauge railway line when a locomotive ran over his leg and practically severed it. It is recalled by several people that, as he was taken away on a hand-propelled, wheeled stretcher, he took out his Bible that he always carried and read it on the way to the factory hospital.

Another accident involving the railway system occurred when an engine driver was taking a standard-gauge locomotive with coal wagons over the weighbridge at walking pace. There was apparently an offending squeak and so the driver got down and walked alongside with an oilcan. Having dealt with the squeak he attempted to climb back onto the running plate and put his foot on the coupling rod. Because the engine had externally fitted Walschaert valve gear, the worst happened and the radius rod came down and sheared off most of his foot.

During the First World War, a Mr Gould lost an eye in the Main Laboratory whilst testing nitroglycerine and he subsequently left the factory and founded Gould's 'Radio House' in Poole. In 1952 a Mr Horlock was painting in a lift shaft in the acids factory when someone operated the lift and the painter was tragically killed. In later years, Mr Horlock's son worked at AMTE, Holton Heath, as a fireman.

The register of accidents for the period shows that a large proportion of accidents were attributed to acid in one way or another, which is hardly surprising since most of the processes in use required acid (sulphuric, nitric or mixed) and these were piped in copious amounts around the site, mainly by way of iron pipes. This may seem surprising to those without a chemistry background; however, the reason that this was possible without corrosion was due to the highly concentrated nature of the liquid. The early stages of manufacture of the acids were also hazardous, as reports of strained backs etc, and persons falling in while cleaning concentrators and retorts, show. Fortunately, the majority of injuries with acids were surprisingly light.

The accident record would not seem particularly bad for a plant of this type and, in case the foregoing has horrified the reader, let us quickly add that by most accounts the RNCF was a very attractive place to work and according to John Gray, who had knowledge of a considerable number of explosive factories, the factory sustained an unusually pleasant atmosphere.

The records indicate that many of the Works Department employees were from Poole and of these most lived in old Poole - in Towngate Street and Strand Street, in Barber's Piles and Globe Lane. Others came from Lagland Street, Bell Lane and Market Street. It is highly probable that many of the production workers also came from this part of the town, most of which has now been redeveloped. These Works Department records also give details of reasons for dismissal and it is interesting, though not surprising perhaps, to note that people were commonly discharged for being in possession of matches in the Danger Area. On one occasion a joiner was suspended for one week after being found in possession of cigarettes.

All employees on starting at the RNCF were provided with a copy of a booklet entitled 'General and Special Rules' which they were expected to be able to produce at any time on demand. All employees were expected to be familiar with the contents of this booklet and the 1936 edition runs to forty pages and no fewer than 137 rules. Fortunately not all the sections were as taxing to the newcomer as paragraphs 4 and 5 in section 1, dealing with definitions, where we learn that:

'4. The word 'clean' as applied to buildings, platforms, trucks, clothing, etc, means that such are so constructed, arranged and maintained as to ensure the absence of grit, dust or other extraneous matter, and are only to be used for or in connection with explosives;

'5. The word 'dirty' may be taken generally to mean the absence of the special conditions enumerated in paragraph 4. It must not be given the ordinary interpretation of 'dirty' as cleanliness in 'dirty' buildings must be maintained.'

As would be expected, there are lengthy paragraphs dealing with the use of tobacco, matches, cigarettes, lighters, etc, and action to be taken in the event of fires or explosions. Rule 24 wisely forbids spitting in the factory whilst Rule 128 insists that clothing and stores that have been in contact with explosives shall be burned on a bright fire, and goes on to remind us that precautions laid down in paragraphs 122, 123, and 124 should be simultaneously observed!

Life must have been somewhat trying for the dedicated smoker employed at RNCF since, in short, he or she could not smoke at all. A merciful management provided a permanent gas jet at the staff entrance to the site in order that withdrawal symptoms could be ended as quickly as possible on leaving at the end of shift. Section five is devoted to thunderstorms, where the responsibility for observing and reporting their approach is fairly and squarely placed on the shoulders of the police. They were to report by telephone the approach of a thunderstorm to the Superintendent himself or, in his absence, the senior officer present. It is particularly interesting to note that the police were not to await the first clap of thunder or the first flash of lightning before making this report. This must have presented something of a challenge to the ablest amongst them, and it must also have been a source of worry to the conscientious. Under such circumstances they no doubt agreed wholeheartedly with Mr Gilbert's well known comment on policemen and their lot! Having received warning of an imminent thunderstorm, the Thunderstorm Warning, consisting of six blasts of four seconds on the meal siren, would be sounded.

In a severe storm all workers in houses handling nitroglycerine/guncotton mixtures were then to leave and go to changing rooms or the nearest building not used for explosives. In the nitroglycerine factory a similar course of action was adopted except that if glycerine was in the course of nitration then some poor soul had to stay on, and could not leave until the senior officer, maybe ensconced in his office, was satisfied that the 'charge' was in a safe condition.

John Gray, who was a nitroglycerine chemist at RNCF during the Second World War, has a vivid recollection of a heavy thunderstorm occurring during August in 1943 when he was in a nitrating house working with one and a half tons of nitroglycerine. He was unable to leave because the separation process was still not complete. It is easy to imagine his anxiety as the storm eventually settled right over the factory with flashes of lightning outside making it look like bonfire night. As it turned out he was safer with the nitroglycerine than was a colleague, 'Holton Heath' Robinson, who had taken shelter against the north-western wall of the fitting shop, for the gable above him was struck and within a second or so he found himself unhurt but surrounded by bricks. Non-matching bricks were used when the gable was repaired and the exact location of this lightning strike can therefore still be seen.

Although it is reasonable to take a light-hearted look at the life of the factory occasionally, the men and women who worked there, management and workforce alike, seem generally to have been of the highest quality. It was unquestionably a first-rate example of how a large number of people could work together amicably and efficiently. The day-to-day operation was an object lesson in team work although it is true that during the wars the commonly shared will to survive was a further and most powerful unifying factor.

It is also true to say that in certain of the more hazardous work areas, interdependence and mutual trust had to be developed literally to the point where one daily placed one's life in the hands of one's colleagues. Such team work was inevitably efficient and productive. The safety of all depended upon a thorough understanding of the rules and also upon their observance. In order to ensure as far as possible that the rules were understood, relevant portions were read by the Foremen to their workers on certain Saturday mornings.

Because there was no tradition of explosives manufacture in the locality it was obviously necessary to bring a certain amount of skilled labour into the area to supply the necessary supervisory effort. Quite a number of these people came from what was then known as the Royal Gunpowder Factory at Waltham Abbey. This was later renamed the Explosives Research and Development Establishment and then the Propellants, Explosives and Rocket Motors Establishment (now closed). Men like Messrs Bugg, Pavely, Austin and Haslar moved house and home to Dorset in order to help get this important part of the British war machine into production. These workers were generally of Foreman grade and they were outstanding men, expert in the manufacture of acids, of guncotton, of nitroglycerine or of cordite and highly respected by workers and management alike.

And so it was that in this part of Dorset the local accent became gradually diluted with that of Essex (given the later influx of population and television to Dorset, it is becoming rare to hear broad Dorset spoken at all!). Certain others were sent to the Royal Gunpowder Factory for training. Cecil Samuel Reyland was an example; he came originally from Crewkerne in Somerset and was a much respected Principal Foreman of nitroglycerine and a great character who retired in 1953.

During and since its use as a propellants factory, Holton Heath has, of course, employed large numbers of Dorset men and women, many of whom spent their complete working life at the site and in some case span two generations with people moving to different jobs and sometimes different employer, as the factory contracted and as other industries moved in. One good example is that of Ted Cake's family. Ted's grandfather was employed as a stoker at the boiler house which drove the Corfe Mullen water pumping station and his father Fred was employed at the RNCF as a bricklayer during the early years. Ted himself started as a plumber in 1953, but before that the factory also employed his three uncles: Reg, the factory chauffeur, Sid, who was a platelayer and Jess, who was engaged in cordite manufacture.

Until his final retirement, Ted remained in Government service and lived in one of the RNCF cottages. His detailed knowledge of the site has proved invaluable on at least one occasion when he was in time to stop bulldozers which were about to cut through a live gas main!

Chapter 2

WATER SUPPLIES & OTHER SERVICES

The scale of the work involved in the construction of the factory can be more readily appreciated when one considers some of the facts and figures which relate to the RNCF. Since an abundance of water is crucial to the satisfactory production of explosives the securing of an adequate supply was of the first importance. This work was commenced in 1915 when the Admiralty established a pumping station at Corfe Mullen to supply the new factory.

The station takes water from the Stour river valley gravels via perforated pipes which feed into a brick-lined well or sump from within the station enclosure which is bounded by black iron railings similar to those surrounding the RNCF itself. The building over the sump housed two magnificent steam-powered horizontal reciprocating pumps. The steam was supplied by a coal-fired Babcock & Wilcox boiler. Sidings were built on to the adjacent Somerset & Dorset railway to supply fuel for the boiler. Coal was transferred from the bunkers into the building on small hand-propelled trucks or skips on narrow-gauge rails.

A small weighbridge, understood now to be with Poole Museums Department, was situated just inside the double-doors through which the fuel passed and as the coal was used so the quantity was recorded on a slate provided for the purpose on an adjacent wall. These engines were each capable of pumping two million gallons in every 24 hours. Each engine had an approximate pump horse power of 67 when pumping at the above rate against a 160 feet head. Both engines could be used simultaneously if the situation demanded it.

The main reservoir, which can be seen on the detailed drawing of

Left: The interior of one half of the main reservoir atop Blackhill Road.

'Area 2' of the site plan, was built on a prominence within the factory known as Black Hill. It was constructed in a semicircular shape and was divided into two quadrants, the eastern one of which was covered over with a reinforced concrete lid. The total capacity of the reservoir was three million gallons and the water was transferred to it by the pumps at Corfe Mullen some six miles away via a 16-inch main. This pipeline passed from the pumping station along the A31 to near St Hubert's Church, then along Knoll Lane and Waterloo Road to Naked Cross. Here the pipeline turned south-west and proceeded via the Limberlost and Lytchett Minster to Holton Heath. Although the reservoir sits on the top of Black Hill, the most prominent point at 116 feet 3 inches above ordnance datum, the station at Corfe Mullen is at 74 feet 9 inches AOD and so the pumps, which in combination were capable of raising water against as much as a 250 feet head, were very well able to cope with the task allotted to them.

There was, in fact, no requirement for the pumping station to be working continually and so daily topping-up operations were undertaken and this was managed as a result of close liaison between factory and pumping station. The procedure was for someone within the main administrative block to climb a short ladder to the roof of the building where, with the aid of binoculars, the level of water in the reservoir could be seen on an externally fitted gauge. The current situation was reported by direct telephone line to a Mr Fuller who was, in the early days, the man responsible for pumping from Corfe Mullen, and the topping-up operation was carried out.

In later years the reservoir was adopted by the Wessex Water Authority and the water-level was measured by an ingenious electrical system employing electromagnetic switches. The equipment was interrogated by 'ringing it up' using the public telephone network and the water-level was transmitted by coded signalling using bells and buzzers. Although the reservoir is currently empty, the feed and distribution pipes are still used to service the local area, the incoming main being bypassed directly to the outgoing.

More recently, just prior to privatisation, one half of the reservoir was fitted with a large steel access door at the base of the hill. This was intended to be used as a provision for an emergency water switching centre, for use in dire emergency, such as a nuclear war or other disaster. It has not been either completed or further used, up to the date of this edition.

In 1934 the western half of the reservoir was covered over. This

Right: This photograph shows intensive work taking place to lay the substantial base for the main reservoir. Note the manual method of brick delivery – also the Foreman closely watching progress!

Right: A view of the main reservoir showing preparation for the columns required to support the enclosing roof structure.

operation required that this quadrant should be drained and that reinforced concrete piers should be built standing on the reservoir floor. These piers supported rolled steel joists which in turn were used as a support for a reinforced concrete lid which was cast over the reservoir. This expanse of concrete was then covered with soil and turf and occasional clumps of heather. An artistic flourish was discovered on the floor of this reservoir where a worker had signed and dated his work. The settlement of lime over the years has unfortunately obscured his name but the date remains.

An empty reservoir of this size is an impressive sight and sound, as the reverberation time is of the order of one minute and a single person singing could imitate a complete congregation!

It was arranged that the factory was as independent of outside influences as was possible. Apart from having to buy in essential raw materials such as coal, pyrites, cotton, paper and a number of other basic commodities, everything was produced with in the factory fence. In fact, during the Second World War in particular, efforts were made to provide duplicates of some essential services. As part of an effort to safeguard water supplies, for example, a diviner was called in early in the war and he found water in the eastern corner of the factory. The diviner claimed that this supply had its source on the other side of the harbour on the Arne peninsula. A pipe was fitted into a borehole 375 feet deep and the valve, attached at the top just below ground level, was never shut off for fear of spoiling the borehole. The borehole continued to supply water at a natural rate of 10,000 gallons per hour and the outflowing stream gave rise to a rich source of fresh, clean watercress until the Wessex Water Authority requested the developers to block it. It was believed that the passage of underground water to the borehole may have led to ground instability on another part of the site.

A pumphouse sat over the borehole and water was pumped into a new reservoir built on a nearby prominence just to the north called Foxhill. Investment in the provision of this alternative water supply amounted to £14,388 with the bulk of the work being undertaken in 1942. This backup supply was never needed in earnest, as the primary supply from Corfe Mullen never failed. This was just as well, since the mineral constituent of the backup supply was only just satisfactory for cordite manufacture. It is worth noting that the pumphouse remained substantially intact until on one occasion Mike Tuck (The NCC warden) was using the building as a counterweight to assist in removing a tree. It was the building which lost the tug of war!

It so happened that the action to cover the western half of the Black Hill reservoir provided the final solution to a problem involving this open expanse of water that had been worrying the Superintendent, Captain Hammond RN, for some years. As a matter of fact he thought he had found a solution in the late twenties but in the event it turned out to be short lived. It all started when, after the change to solventless cordite, thousands of wooden trays, formerly used in the drying process, had become redundant and were given away to any who could find a use for them. These trays were thirty-nine inches long by twenty-one inches, shallow, and fitted with a diagonally slatted bottom to allow the free passage of warm air in the drying process. They were well-made and each one was held together with precisely seventy-six brass screws. The trays apparently appeared like a rash all over the neighbourhood as they were very popular in the construction of fencing and ornamental garden gates. Such a structure was to be seen in the Holton Heath area in 1981, still looking in reasonable condition over fifty years later.

The real point, however, was that the Superintendent had been offended for some time by the audacity of flocks of seagulls that frequented the uncovered western half of the reservoir, flying over it, swimming on it, fouling it even and generally making the place look untidy. The availability of the drying trays seemed like the answer to a prayer to the Superintendent who, though land-based, was obviously still a practical sailor at heart, for he decided to float

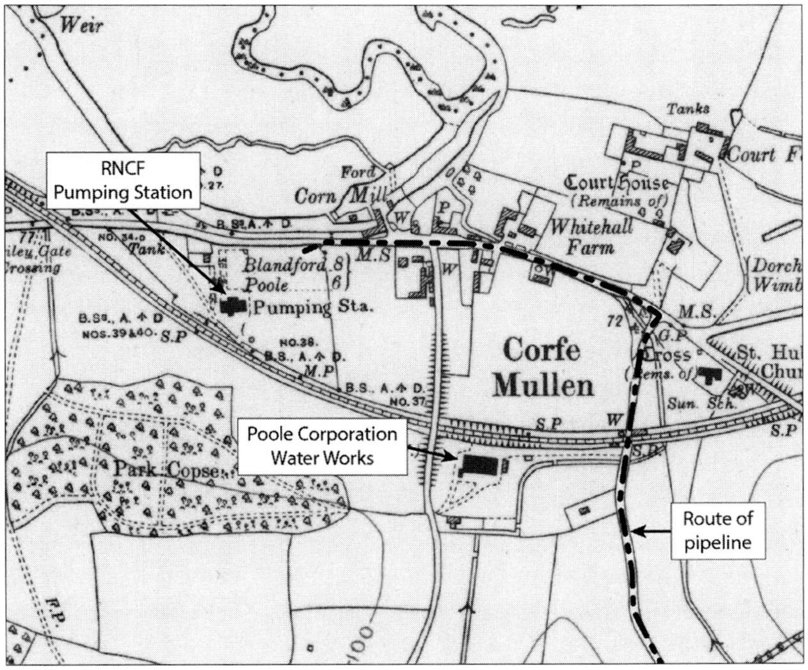

Above right: Map showing the location of the pumping station and pipeline route.

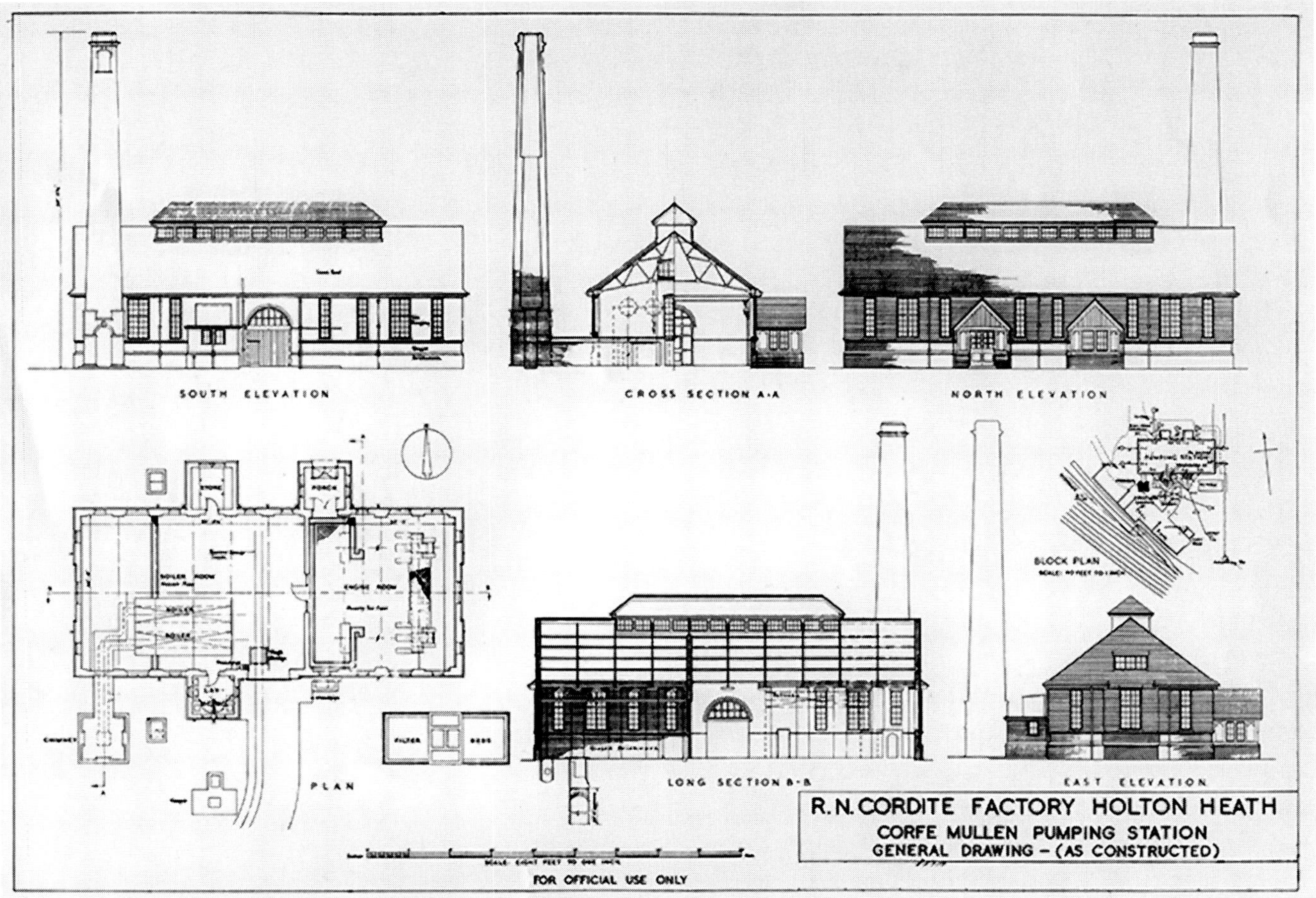

Right: A reproduction of the construction plans for the pumping station.

hundreds of these wooden trays on the water with the conviction that these would render the factory less attractive to the sea birds.

Whether the ploy would have been successful will never be known for there were unacceptable side effects in that residents in the police houses, and other RNCF dwellings whose domestic water supply came from the factory reservoir, started to complain of headaches and all the trays had to be removed. The explanation for this was to be found in the presence within the wood of minute amounts of nitroglycerine which had been absorbed over the years from the drying cordite. Nitroglycerine is the common or trivial name for glyceryl trinitrate and, apart from being a very powerful explosive, this material has a number of interesting physiological effects when it is absorbed into the body via the skin or mucous membrane. Exposure to the liquid or the vapour (which has a smell similar to that of a recently extinguished paraffin lamp) causes severe headaches and throbbing of the head, and these were the symptoms which were experienced by the local residents.

Severe headache was, in fact, yet another occupational hazard that workers had to live with until a tolerance was gradually developed as a result of repeated exposure. Some workers were never able to tolerate nitroglycerine however and had, with some regret, to be employed in other parts of the factory – the regret being caused by the fact that higher pay was available for working with nitroglycerine.

However, nothing is all bad and, used medicinally in controlled doses, commonly as 500 microgram tablets in a mannitol basis administered sublingually, nitroglycerine is a valuable drug for the control of angina. When used in this way it has the effect of tending to restore the blood flow to the heart muscles, thus relieving the severe chest pains and pain in the left arm which are symptoms characteristic of this condition. Ted Terrell recalled another medicinal use for cordite – not recommended by the medical profession – which dates from the conscription period of the First World War. It seems that it was common practice for those 'in the know' to chew cordite. This produced an abnormally fast heartbeat and lead to the individual being rejected by the medical board which periodically came to the factory.

The factory had a boiler house, a generating station and even its own gas works. The boiler house was a massive affair (which stood for a considerable time, with subsequent uses, alongside the railway line) about 340 feet long and equipped with twenty-four large coal-fired boilers which were used to supply live and dead steam through insulated pipe lines to all parts of the factory. Pipelines ran to enormous lengths and the problem of heat loss must have been severe, but the asbestos insulation was effective and steam was available anywhere on site.

A small oil-fired boiler house was used to supply additional steam for the decontamination building associated with the 'Z' cordite

Below: The exterior of the main boiler house photographed in 1987. The tubes in the foreground are spun nylon made by Fibaflow Ltd, the occupiers at the time, which formed the core of MRI scanners. The coal entry point was at the far end of the building, close to Holton Heath station.

Above: The interior of the main boiler house, close to the Wareham – Poole railway line. These boilers generated steam for the entire factory and were fed on a continuous basis from an overhead coal conveyor system.

pressing ranges. This was possibly installed to meet a shortfall or as backup. The first edition of this book wrongly ascribes this boiler as being for charging the narrow-gauge 'fireless' locomotives. Later information clearly shows that these were, in fact, charged in the engineering section, closer to the main boiler house. This small boiler house was for some reason spared when bulldozers were levelling a part of the site now owned by British Land Company and it can presently be seen standing quite alone in a sea of gorse close to Blackhill Road.

In later years when the full capacity of the generating station was not required, excess power was fed into the national grid. The tar produced at the gas works was mixed with sodium silicate to give a concoction known as Clarke's Enamel, presumably after its inventor, which was used as a preservative on the steel fence which surrounded the factory. It was popular at the time to claim that this product presented a greater corrosive hazard to the fence than did exposure to the weather!

Simultaneously with the building operations an approximately three-and-a-quarter-mile-long 12 feet high so-called 'unclimbable' steel fence was thrown up around the factory site and a sewage plant was built just to the south-east of the main railway line. Many other essential support facilities were also provided including a whole range of stores and a variety of workshops. Several laboratories were built, two of the larger ones being the Main Laboratory which faced the General Office in the western corner of the factory and the Acetone Laboratory, which later became the Physics Laboratory, which was just to the north-west of the main gates. A number of messes and canteens were provided and latrines were widely distributed over the entire factory area.

A further laboratory, 'X3', was located within the Danger Area at the east end of Knights Bridge. This initially dealt exclusively with nitroglycerine. Peggy Walters was relief assistant for X3 and recalls 'We all had a healthy respect for nitroglycerine, not only for the obvious safety reasons, but any spilt on the hands would result in the dreaded nitroglycerine head – a real throbbing headache. The bench we worked at had a line painted several inches from the edge and we had to work inside the line to avoid any accidental splashing on the floor.'

All these services were in support of the main manufacturing processes which were carried on within the boundary fence. At this point it might be as well to point out that, although the whole was called the Royal Naval Cordite Factory, geared primarily to the production of cordite, there were a number of sub-factories, as it were, devoted to the production of, for example, acids, acetone, guncotton, nitroglycerine and – much later – picrite. Therefore, throughout this account it is only when 'the factory' or 'the RNCF' is mentioned that a reference is being made to the entire complex within the boundary fence.

A further early requirement in the construction of the factory was for sidings to be provided to lead from the main railway into the factory area in order that equipment and building materials could be delivered quickly and easily to the site. Later, these sidings would play an important part in transporting raw materials into and finished products out of the factory. Mr W L Brown, who died at the grand old age of 90 and who was on the railways all his working life, clearly remembered dropping the first load of points at the remote Holton Heath site in 1915. At this time he was fireman on the Beyer & Peacock locomotive which drew the train carrying the points. In fact, the construction of the standard-gauge network was started in the March of that year and a signal box was opened to control the sidings on 10th May 1915. Harry Bugg remembered well a temporary station with a single platform on the up side constructed just to the east of the boiler house. In those early days, Drummond rail motors were used for some trains from Poole, Christchurch and Bournemouth West. The present station opened on Monday, 3rd April 1916 but it was not until Monday, 14th July 1924, soon after the changeover from the L&SWR to the Southern Railway, that it became available for the use of the public at large. The station, although still in use, has of course seen much reduction of facilities in later years, including the removal of buildings.

Road transport at that time would have been comparatively inefficient and would have depended partly upon horses and steam traction. Nevertheless, the roads were also used as was made very clear in a classic case of 'careless talk' when a lady cyclist on holiday wrote a postcard to a Miss Whyte in Bournemouth. On the card, which was postmarked 19th April 1915, the holidaymaker was contrasting the serenity and beauty of East Lulworth with the chaos and terrible state of the roads near Wareham where 'they are building an enormous factory for munitions with engines going to and fro all the time. I passed two punctured motors and two punctured cyclists.' One cannot but wonder at what the lady would make of the same road today.

A further extension of the standard-gauge railway was taken eastwards to beyond the limit of the proposed factory area where it

Above: The electricity generating plant, located at the rear of the boiler house. Surplus electricity from this plant was supplied to the national grid.

Above: The view inside the machine shop where such items as cordite press dies were fabricated. Note the use of overhead pulleys and belts for power, rather than the individual electric motors which are the norm today.

was then carried south over the main Waterloo to Weymouth line and thence to a specially built jetty at Rocklea where the harbour had been suitably dredged to allow access from the main channel. The jetty was equipped with a flagstaff at the end, two pillar cranes, a rail-mounted crane and office accommodation, whilst nearby on the shore was a store, a mess room and a boat house with a slipway.

Mr Tom Morgan (not the superintendent of the same name) compiled some further information regarding the jetty and pointed out that this was rented out to the firm of Salway-Morgan (marine and general engineers) in the 1950s and it was then used for fitting out small craft such as privately purchased, ex-Admiralty motor torpedo boats. Access for the workers was normally via a pinnace departing from the Fish Shambles steps. Anyone missing the boat had to get to Hamworthy Junction station and, after getting out of sight of the station, risk a walk along the main line railway and access the jetty by the Admiralty rail bridge which crossed the main line. The jetty proved to be a popular location for fishermen until it was finally broken up and removed.

Chapter 3

TRANSPORT SYSTEMS

The factory used and produced vast quantities of a whole range of materials and, because of its large size, they all had to be transported over quite significant distances. Consequently, the transport systems employed were appropriately impressive, not only in size but also in their variety and their novelty. Reference has already been made to the standard-gauge railway system which existed and this ultimately amounted to something like five miles of track within the factory. Normally two small saddle tank locomotives were in use within the factory at any one time and a steam locomotive shed was erected for their use in the southern corner of the site. The standard-gauge railway system was used to transport stores, equipment and raw materials into the factory from the main national network and also to deliver the factory's products to the main line sidings or to transport material to and from the Rocklea Jetty. The main general purpose transport network within the factory was the 2 feet 6 inch narrow-gauge railway system which penetrated to virtually every corner, serving all the buildings associated with the production of

Below: The standard-gauge locomotive on the left is Yard No.1596, seen at Holton Heath on 17th July 1959. This 0-4-0 ST was built by W G Bagnall of Stafford as works number 2596 in 1938. This 27 ton engine had 14 x 22 inch cylinders and 3 foot 6 inch wheels. It was sold for scrap around 1961.

cordite and many others as well. In the early days, because of the nature of the factory's products, special fireless locomotives were used over the factory's fourteen miles or so of track. These depended upon a charge of high pressure steam which was stored in an insulated chamber. Two recharging stations existed within the factory. These 'fireless' locos, also known as 'Fowlers Ghosts', guaranteed a source of spark-free traction and were capable of working a half shift between charges. As an alternative to this, battery driven locomotives were frequently used. These were commonly of two types manufactured by Greenwood & Batley of Leeds. The first type was merely an electric railcar with the driver standing on an open platform at the front and the load being carried on a larger platform at the rear. The second type provided a covered-in cab for the driver and was designed to pull a rake of trucks. These battery operated locomotives also required charging of course and stations were provided for this purpose outside the Danger Area fence. Locomotives of this kind sometimes pulled trucks modified to carry passengers, not in the normal course of factory activities but when groups of important visitors attended the factory or on the occasion of social functions such as the annual summer flower shows when workers and their families were allowed into the factory.

Ruston & Procter Ltd of Lincoln produced a locomotive during the First World War, apparently suitable for use in explosive factories, which was fuelled by paraffin. At least one of these 2 feet 6 inch 10 Horse Power units (No. 52124) was received from the manufacturer

Below: This illustration of the railway system in the acids factory area, gives some idea of the complex standard-gauge rail network within the factory.

in 1918. We are not sure whether or not it was used within the Danger Area. Drivers of such locomotives were expected to be familiar with fifty-five rules, listed in the booklet, *Royal Naval Cordite Factory General and Special Rules* which applied specifically to them.

In his excellent 'Rails to Poole Harbour', Colin Stone claims that at its peak, as many as thirty-four locomotives were employed on this narrow-gauge railway. Mention is also made of a spur off this system which passed southwards under the main L&SWR track to the harbour. One of us well remembers finding evidence, some fifty years ago, of this extension that led to the margins of the harbour and large, overgrown heaps of what was understood to be waste material from the acids factory. Also, cut-offs and failed pressings of vinyl attested to the later period when Decca records were on site.

Having been transported from the factory to Rocklea Jetty via standard-gauge rail, the cordite was then taken in barges to destinations such as the RN Armament Depot at Priddy's Hard, near Portsmouth Dockyard, and to Upnor Castle on the Medway, near the dockyard at Chatham. There could be no greater authority on the transportation of the factory's main product by sea than Eddie Grist who was personally involved from as early as 1930 when he joined the Priddy's Hard RN Armament Vessels Department in the October of that year as a deck boy. He recalled sailing in *Emily*, on his first voyage, to Rocklea Jetty for a cargo of cordite.

Below: Carrying cordite to the press houses on the narrow-gauge railway system. Note the fire-resistant 'black lasting' clothing and cap. The coats and trousers had no pockets or buttons – tapes were used to fasten the clothing. The curved shape of a press house roof may be seen behind the tarpaulin.

At one time the *Emily* and *Henry Gordon* plied between Rocklea and Priddy's Hard whilst the *Egypt* and the *Alfred* operated to and from Upnor. The *Maria* and the *Betsy* were also involved carrying cargoes of cordite from Rocklea. The barges sailed to the Stakes, a mooring in Poole Harbour, and from there they were towed by local fishermen to Rocklea. This was necessary because they were incapable of tacking and weaving through the tortuous channel. The vessels were double-rigged, flat-bottomed and fitted with leeboards to assist sailing qualities. They were crewed by a mate, two able seamen and a boy. Later on, these Thames barges were replaced by steam-propelled barges, including the *Bowstring* and the *Lanyard* and by steel dumb lighters. Most of these vessels carried about 100 tons of cordite. Transport by sea was discontinued in 1938, after which the use of rail transport was deemed to be rather safer from aerial attack. Eddie finally retired after a 46-year seafaring career culminating in his appointment as Shipping Master at HM Naval Base, Rosyth and subsequent retirement at Portland Naval Base.

Other transport methods included the use of pipelines, employed in the usual way for water, gas and sewage but also for live steam which was used throughout the factory for space heating, drying and for the provision of heat in certain of the manufacturing processes. Large diameter asbestos-insulated pipes were used for this purpose. Hot water was generated locally at groups of buildings by passing the steam through heat exchangers, or calorifiers, which were sited close to the buildings they were intended to serve. John Gray recalls an incident involving a calorifier which had developed a leak. As the steam was at a higher pressure than the cold water supply, steam was driven into a lavatory cistern. A lady 'pulling the chain' was rather alarmed when this resulted in much turmoil, noise and steam issuing from the WC pan!

Mixed acids were transported from the acids factory to the Nitroglycerine Hill area by means of a pipeline carried on a gantry. In later years, after 1927, when a newer form of cordite (SC) was being produced, nitrocellulose was carried as a suspension in water in a pipeline from the guncotton factory to the mixing houses at the foot of the Nitroglycerine Hill for incorporation with nitroglycerine in the earliest stages of cordite manufacture.

The transport of liquid nitroglycerine was naturally kept to a minimum, but it did have to be moved from one process to the next, and this was achieved by causing the liquid to flow under the influence of gravity from one house to another along open gutters made of lead and fitted with warm water jackets to avoid the possibility of it freezing. The gutters were butt welded to provide a satisfactorily smooth surface. The fall of such gutters was slight but not less than 1 in 65 and they were entirely free of sharp turns or of crevices or recesses in which the liquid could lodge. Pipes were entirely unsuitable for such purposes because of the impossibility of inspecting or cleaning them properly.

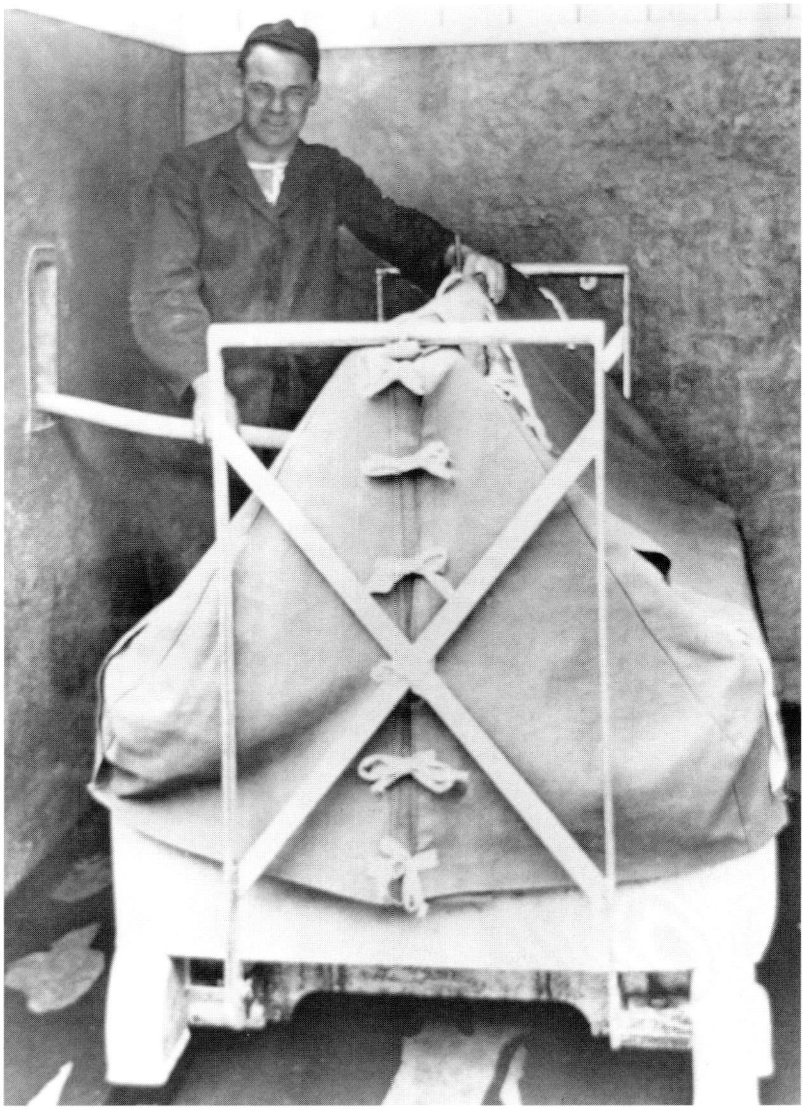

Above: Bert Fricker in charge of a hand-propelled rubber-wheeled nitroglycerine trolley, which is being filled at a washing house, prior to being pushed along the grooved asphalt runway to a burette house.

The gutters were protected by means of canvas coverings fixed along one edge and laced down on the other so as to facilitate cleaning between the passage of charges. This was achieved by folding back

Above: Sid Walker from Wareham (sitting) with a workmate on a battery-driven locomotive hauling two trucks laden with packed cordite. This load was probably destined for the magazines ('M' houses). Tractor No.1 of this type was reserved for moving paste trucks only and was especially low-geared because of the hazardous nature of the product.
Note the small building at the rear, which is one of many small electricity sub-stations and carries the adjacent press house number: 'P10'.

Above: Fireless locomotive No. AB 1475 built by Andrew Barclay & Company Ltd at the Caledonia works, Kilmarnock. This was one of the narrow-gauge locomotives sometimes known as 'Fowler's Ghosts' employed at the RNCF.

the canvas and wiping along the entire length of gutter between houses with a cotton material known as 'flicker cloth' starting at the higher point. Nitroglycerine factories are often built on rising ground in order to provide a fall from each house to the next without having to make the first in the series of great height. Hence reference is often made to 'Nitroglycerine Hills' even in more recent times when modern techniques have overcome the need for a natural slope. By the same token, workers in the nitroglycerine factories are known as 'Nitroglycerine Hillmen' in factory jargon.

At Holton Heath this simple transport system for liquid nitroglycerine was used between each house involved in the manufacture and refinement of the explosive until the Nitroglycerine Hill explosion of 1931, when a nitroglycerine-wetted gutter caused the explosion to spread to adjacent houses. After that time a mechanical transport scheme was introduced which involved the use of a hand propelled trolley with rubber wheels which fitted in parallel grooves formed in a smooth asphalt track, passing from the wash houses to the burettes adjacent to the mixing houses. Gutters were still retained for passing the nitroglycerine to the wash houses and for carrying away wash waters to the wash water settling house.

It is interesting to note that road access played very little part in the site transport system and the railway was provided mainly for the movement of material. Staff whose work took them to the farthest corners of the factory must certainly have received their share of exercise. Some of the old track bed was later converted to road use and the present Blackhill Road follows part of tracks Nos. 3 and 10.

Chapter 4

THE FACTORY BUILDINGS

Buildings within the factory fence were basically of two kinds; those designed for the manufacture or storage of explosives and those built for other purposes. Those built for the production or containment of explosives were called 'Danger Buildings' and were all situated within the Danger Area fence. Each house within the Danger Area concerned with the production of explosives was identified by means of a letter or letters and, unless it was unique, by a number as well. The key started with 'AB' which was indicative of nitration and separation of nitroglycerine and 'ABC' which included washing as well as the foregoing, whilst 'C' houses were employed exclusively for the washing of nitroglycerine. This continued with 'D' for wash water settling houses. 'E' was reserved for paste mixing, where nitroglycerine and guncotton were first mixed, and 'F' was the drying house for the paste received from the 'E' house. This logical sequence followed the same order in which the manufacturing processes followed each other with 'G' being the rolling houses and 'H' the charge forming houses. After this, however, the sequence seems to have been abandoned, since the next stage in cordite manufacture is pressing and the press houses are designated 'P', while picrite factory buildings are designated 'N'. This anomaly is probably explained by the evolution of various types of cordite, a number of which were manufactured at RNCF. The current key to the use of these letters appeared, revised as necessary, in each edition of the *General & Special Rules for the RNCF*, the last of which appeared in 1936.

The danger buildings, which numbered over 120 even in the early days, included those grouped within the guncotton factory, the Nitroglycerine Hill buildings, the tetryl factory, guncotton drying stove, acetone recovery stoves, the ranges where cordite was pressed, blending houses and magazines.

Perhaps the most reliable source of information on danger buildings relevant to the early days of the RNCF is HM Inspectors of Explosives Annual Report for 1909. That recommendations made at that time were still appropriate when the factory was built in 1915 is supported by the fact that this report was quoted in the second edition of Arthur Marshall's definitive work *Explosives* which appeared in 1917, at least one year after the completion of the factory. It is certainly true that all the information which has come to light about these buildings has been entirely compatible with the recommendations which appeared in the 1909 report.

HM Inspectors of Explosives stated that 'In working buildings where all the explosive is removed on completion of the day's work, the more flimsy the construction the better. In view of the distances to which heavy masses of debris have been thrown by explosions, the soundness of the principle is obvious. The actual material recommended is matchboarding, corrugated iron or sheets of light, fire-resisting material. For roofing, slates and tiles are dangerous, and tarred felt carries fire, so that only corrugated iron and fireproof sheets remain, and there are objections to these; in fact a good roofing material for working buildings containing explosives is much wanted.'

At Holton Heath the Nitroglycerine Hill buildings, and a large number of others in the Danger Area, had their walls and roofs of wood. The roofs were covered with a rubberoid material which, though flammable, was readily extinguished when projected through the air and it was because of this that the Inspectors had given the material their blessing in their Annual Report of 1910.

The 1909 report went on to say that 'the lining of a danger building need not necessarily be of wood. In fact in dusty buildings a lining which can be effectively washed down by means of a hose has many advantages.' As it happened, the nitroglycerine production houses were lined with wood but the guncotton drying stoves were lined with zinc to facilitate cleaning. The picrite factory, which was built much later, just prior to the Second World War, used white glazed bricks on inside walls but by this time, partly because of the greater threat of aerial attack, a different set of constructional principles was in operation. During the Second World War, many of the drying stoves had concrete roofs installed to mitigate against bomb strikes. This was achieved by using the existing roof as a form and pouring concrete over.

'The floors,' said the Inspectors, 'should be of soft material such as wood or asphalt, free from grit; care being taken to fill up all interstices and crevices in which explosive dust may accumulate.

Doors should open outwards and, while anyone is in the building, should only be secured in such a manner as to be opened easily by a push. There should be plenty of them, and so far as practicable they should face towards the path by which a work-person would naturally endeavour to escape.' Asphalt was commonly used for floors at the RNCF and doors in danger buildings seem to have met the Inspectors' requirements insofar as they were secured in the approved fashion and there were plenty of them in the right places. Doors did open outwards and they were secured by a shaped wooden handle which extended beyond the edge of the door and which clipped

Above: One of the asphalt runway tracks leading to a burette house, shortly after completion. The nitroglycerine was delivered to the burette house by truck and then sprayed (by siphonage) on to a liquid nitrocellulose pulp. The mixture then passed by gravity down to the mixing house at right. In the distance may be seen the acids factory complex.

into a gunmetal fitting attached to the jamb. When the doors were open they could be secured in this position by means of gunmetal hooks. When closed, double doors were secured on the inside with gunmetal bolts. In order that the moving parts of such door furniture could not be removed they were permanently attached by leather straps. Doors could be locked on the outside by means of padlocks which were also made of gunmetal or brass. The use of these metals reduced the risk of sparks being struck and, for the same reason, nails and screws used in the construction of the danger buildings were also made from these materials.

Another very important safety measure required that danger buildings should be at certain prescribed minimum distances from other structures or places where people were working. The Inspectors published a table of such distances which were related to particular kinds of buildings and the amount and nature of the explosives held in them. These distances could be halved if adequate earth mounds or traverses were provided, or even reduced to one quarter if a natural hill intervened. There was, however, one notable exception to all these provisions and that was in the case of 'a palace or house of residence of His Majesty, his heirs and successors', where no danger building whatever was allowed to be within two miles. It is also interesting to note that in the case of a 'theatre or other building where people are accustomed to congregate' then certain danger buildings need to be no more than fifty yards away.

Mounds or traverses were erected around danger buildings to prevent an explosive wave travelling horizontally to other danger buildings nearby. At RNCF, earth traverses were perhaps twenty to thirty feet thick at the base and reached to the height of the roof. Access was commonly, though not exclusively, via short tunnels through the earthworks. Pedestrian access was sometimes arranged via wooden stairways and mechanical transport sometimes gained access through gaps in the earthworks.

Artificial lighting at RNCF in the nitroglycerine houses, and in certain other danger buildings, was provided by electric lamps which conformed to legal requirements by being mounted outside the buildings and arranged in such a way that the light shone through the windows. All the wiring, connections and switch gear was also outside the buildings. Elsewhere in danger buildings, if this technique was not employed, the lamps were situated within the buildings but they were then always protected by thick glass covers so that a blow sufficient to break the glass would inevitably also break the filament. An example of this was the *Wigan* light fitting which was very popular at the time. All wiring was encased in steel conduit such that if a short circuit should occur no external ignition would result. These precautions were the forerunners of the 'intrinsic safety' techniques now commonly used in explosive atmospheres.

Windows and skylights were commonly of wire reinforced glass and it was recommended that, with the exception of those facing north, they should be dulled in some way in order that the deleterious effects of direct sunlight on the explosives should be avoided. Frosted glass was often used at Holton Heath and in more recent years windows were made from thick glass 'bricks'. Each danger building had a wooden board attached in a prominent position on its exterior, usually above a door, upon which the maximum explosives content in pounds and the number of workers allowed was shown. In no circumstances were these limits to be exceeded. On some of the entrance walls to those drying stoves and magazines which still stand, pencilled addition sums can be seen where the packer has checked the total load.

All danger buildings were fitted with lightning conductors and some, depending upon their size, were equipped with two or more. Even at that time, opinion was divided on their effectiveness. Some, including Marshall, thought it unreasonable to expect lightning to follow the tortuous route of the copper conductor strip as it passed from some point near the centre of the roof to earth. His view was that lightning commonly takes short cuts and that such fittings were more likely to conduct lightning into a building than away from it. In particular, nitroglycerine nitrating houses, such as the AB and ABC houses at RNCF, were especially liable to be struck, partly because they were usually higher than the rest of the factory and partly because they contain large masses of metal. This metal, mainly in the form of lead, which was used for all the processing vessels, could give rise to a spark induced by a flash striking near the building. For this reason all the metalwork was connected together and to earth by conductors. However, although some of the danger buildings fairly bristled with lightning conductors there do not seem to have been any serious accidents or damage recorded as a result of strikes.

John Gray recalls being told by a Hillman of a lightning strike on building 'C1' when the conductor proved to be effective but the enormous current melted the copper strip and the mark of burning could be seen all along its course over the wooden structure. There was no other effect on either the nitroglycerine or on the building.

Architecturally, the most satisfying group of buildings within the factory was that including the General Offices which surrounded the courtyard inside the office gates which are most conveniently approached by road (Station Road) leading south-east from the main A351. This entrance features four tile-capped brick piers, with the two widely spaced central piers supporting a massive pair of black iron gates. The space between the first and second of these piers as one approaches the entrance is permanently fenced in and serves merely

Above: A fine view from the main reservoir towards Poole. In the foreground is the Charge House feeding the 'ABC' house. In the distance is a group of large buildings comprising the acids factory. Judging from the bare earthwork, this photograph probably dates from 1917.

Above: A view of the main laboratory where routine inspection and quality control was carried out. Further checks of the product were also made by the Inspectorate of Naval Ordnance in a separate building.

Above: A second view of the main laboratory. Mrs Peggy Walters (then Peggy Barber) is at the left of the picture and informs us that this was a second shift around 1944-5. The girl to the right is Brenda Hughes (later to become Mrs Diffey) and the gentleman by the door is Mr David Butler. It appears that Brenda usually worked from the other side of the bench, but had to move around for the benefit of the photographer!

to balance the similarly spaced third and fourth piers which support a narrower iron gateway for the use of pedestrians. The pavement leading through this gate passes immediately in front of the brick built police box and from this point the whole courtyard can be seen. The area is dominated by a large brick-built building facing north-west into the courtyard and which housed the main General Offices of the factory. The building is of two stories with attic accommodation and dormer windows. The pleasing red brick used to build this office was used in all the buildings surrounding the courtyard.

In the centre is the flagstaff at the foot of which was situated a figurehead, facing the gates. The first figurehead to be installed was that from HMS *Constance*. This choice was a considered one and was made because she was one of the first, if not the first, naval vessel to use cordite. This figurehead, carved from white pine planks jointed and glued together, ultimately succumbed to the effects of wind and weather in the early 1950s when she was ceremonially 'cremated' on the cordite burning ground. She was replaced in about 1952 by the figurehead from HMS *Rattlesnake*, and this remained until closure, on its concrete plinth, flanked on either side by a bronze mortar.

After the site closed and DERA staff moved to other locations, this figurehead was superbly restored by Mr A E Mechen (of Netley Marsh, Southampton) who had salvaged many similar figureheads for many customers including the Royal Navy. Confirmation of the hopes of an enlightened management to restore this figurehead was provided after one of us viewed *Rattlesnake* after the crudely modelled and painted glass reinforced plastic (fibreglass) cladding had been removed to expose the original pine carved structure which, though quite rotten on the inside, had retained a sound and beautifully modelled surface, probably as a result of additives in the wet lay-up fibreglass providing antifungal/antimicrobial protection. This magnificent figurehead now resides in the concourse at QinetiQ at Farnborough.

The inscription on a wooden plaque below the figurehead reads as follows:

'HMS *Rattlesnake* 6th rate Frigate 26 guns. Length of gun deck 113 feet 9½ inches. Breadth (extreme) 31 feet 7½ inches. Depth in hold 8 feet 9¼ inches. 503 tons. 175 men. Built at Chatham and launched 1822. Was employed surveying under Captain Owen Stanley 1847-50 and as Arctic relief ship 1853-6. Broken up at Chatham: 1859-60.'

This rather bald summary appears to be correct in every respect except that she was fitted with 28 guns rather than with 26. This ship was the fourth to bear the name, but since 1860 there have been four

Below: An external view of the administration building taken shortly after all MoD staff had moved out – sadly taking with them the figurehead as witnessed by the empty plinth.

other *Rattlesnakes* in the Royal Navy. Her immediate successor was a corvette (1861-1875), the first of the name to be steam-propelled, and this ship was followed, successively, by a torpedo gun boat (1886-1906), a torpedo boat destroyer (1910-1921) and lastly by a minesweeper built in 1943.

Facing the office block across the courtyard is the second largest building in the group, this time a single-storey laboratory block with dormer-windowed attic accommodation. Each of these main buildings has large brass-handled double hardwood doors leading into entrance halls from the courtyard. The flagstaff is sited in the centre of a small lawn in such a way that it bisects the straight line between the two doors facing each other across the court. On the edge of the court immediately opposite the office gates and also in a straight line with the flagstaff lies a smaller third building also of single-storey construction.

The layout of this area as so far described has a satisfying symmetry, but monotony, on the other hand is avoided and additional interest is provided by a number of other buildings, the most significant of which is the Office Mess Room situated just inside the gates to the south-east. Between this building and the General Office block, and set back a little further from the court is another smaller building surrounded on three sides by a lawned area. These two latter buildings are of similar construction with hipped roofs. To the north-west of the main laboratory block and on a lawned area of lower ground are sited three very small single-roomed buildings serving as stores for the laboratories. These are all built in matching brick and have gable-ended roofs with similar tiles to those used in all the other buildings.

These completed an attractive group which owed much to pleasing proportions in both the buildings and the space between them coupled with variety in size and in shape, with the whole bound together by the common use of attractive brick and tile. The appearance of this area must have improved with the passage of time as the lawns became more established and trees grew, but gradually, as pressure increased on available space, various additions and modifications were made. In this way the attractiveness of this area passed through a high point, probably in the 1930s, because

Left: An aerial view, not of the factory but a large and very accurate scale model made during the Second World War by R W Hucklesby and staff of the Works Drawing Office. This model was stored for some years until being returned to Holton Heath for conservation work. When the MoD left the site the model returned to the *Explosion!* Museum at Priddy's Hard.

all subsequent changes were architecturally disastrous with the use of clashing brick and the erection of wooden buildings with felted roofs and the growing use of asbestos cement sheet in temporary buildings. Because it was a factory with a deadly serious purpose it may well be irrelevant to question building styles but it is interesting to reflect on the concern that the architects had for the appearance of this part of the establishment when it was built in wartime in 1915.

Situated beyond the courtyard to the north-east was a large wooden bungalow surrounded by a verandah in the Colonial style. This was originally built to house the factory manager and was situated in a garden equipped with a greenhouse and bordered by a screen of trees. An ornamental fish pond survived into the 1990s although the bungalow itself, after serving for a time as office accommodation for the Inspectorate of Naval Ordnance, was finally used after the closing of the factory as a games hall, with table tennis and billiards facilities. The bungalow was demolished in the early sixties and the site was then used as a car park.

The other entrance to the factory is on the same side of the road and some 200 yards south towards Holton Heath railway station. At this point a similar set of iron gates some 12 feet high are supported between similar brick piers. Just inside these gates on the right is a single-storey building which then housed police, timekeepers and pay offices. On the left-hand side was the factory hospital which is also of single-storey construction and brick built.

The rest of the buildings within the factory, but outside the Danger Area fence, are mainly of brick construction but many of the smaller ones were, like the danger buildings themselves, rather unusual in that felt covered wooden roofs with Belfast trusses were used. Belfast trussed roofs may be recognised from the outside by their gently curving shape.

In addition to the accommodation provided for the police, and the MDS's bungalow already described, a number of cottages were built for the use of staff. Eighteen of these were built of yellow brick in the western corner of the site just outside the factory fence. They included fourteen semi-detached three-bedroom cottages and three detached four-bedroom properties. Although the relatively populous areas of Wareham and Poole were close in terms of short railway journeys there must obviously have been a need to have certain key staff immediately to hand and the cottages met those needs.

The last staff house to be built was erected in 1927 and is described on subsequent site plans as the 'Superintendent's Residence'. This building, which still stands and is now known as Holton Heath House, accommodated for many years the Commanding Officer of the Royal Marines establishment at Poole. The house, which is in brick and has five bedrooms, stands in grounds of just over one and a half acres and had its own tennis court in the southern corner.

After the Great War in 1919 a separate plot of land, squarish in shape and on the opposite side of the road from the factory, was cleared and a series of laboratories and climatic huts were constructed for the use of the Inspectorate of Naval Ordnance. The larger buildings were constructed in a style similar to that used in the factory proper whilst for the climatic huts, in which cordite was to be tested under various environmental conditions, much use was made of wood and asbestos cement sheeting and blue asbestos insulation. As in the factory, earth traverses were thrown up around those buildings which were to contain explosives. No doubt because of their laboratory-based origins, some of these buildings – together with modern additions – were later employed by the Admiralty Materials Laboratory and its successors for various chemistry-based research projects right up until the closure of the site in the mid-nineties.

Left: A view towards the 'main' gates. The more imposing entrance which accessed the administration area, was known as the Office Gates. This picture shows at left, the police office and at right, the hospital. In later years these fine buildings were somewhat spoiled by the imposition of flat roofs and the arches in front of the police office were built across – the brickwork being covered with paint.

Chapter 5

ACETONE & CHAIM WEIZMANN

The manufacture of cordite in the First World War depended upon the solvent action of acetone to assist in the incorporation of nitroglycerine with guncotton. The difficulty was that there was a world shortage of this material and so it became an important priority to ensure that the factory had a reliable supply of its own. It was in this connection that a very famous Zionist, Dr Chaim Weizmann, became associated with the RNCF.

Dr Weizmann was a biochemist of international repute and was the first reader in this subject at the University of Manchester. At the end of August 1914, he had freely offered his fermentation process for the production of acetone to the War Office. This action was in response to a request from that Department asking every scientist possessed of discoveries of military value to report them. Initially, there was no response from the authorities and it was not until one

Below: A general view of the acetone factory, shortly after completion. The huge concrete fermentation vessels can be seen, partly outside of the building.

day in March 1916 that he was invited to visit the Admiralty at the request of Sir Frederick Nathan, who was at that time the head of the Powder Department. Nathan explained that there was a serious shortage of acetone and went on to say that if the deficiency could not be made good, then extensive changes would have to be made to naval guns.

Weizmann was thereupon invited to work on the problem which involved putting up an acetone fermentation plant in London. During this time he was still carrying on with his teaching duties in Manchester. Ultimately he found the strain too great, involving as it did, travelling between London and Manchester by night to save time and he became more or less fully employed by the Government.

He was finally brought into the presence of the First Lord of the Admiralty, Winston Churchill, who, according to Weizmann in his autobiography *Trial and Error*, greeted him briskly with the words 'Well Dr Weizmann, we need thirty-thousand tons of acetone. Can you make it?'Terrified at this lordly request he answered 'So far I have succeeded in making a few hundred cubic centimetres of acetone at a time by the fermentation process. I do my work in a laboratory. I am not a technician, I am only a research chemist. But if I were somehow able to produce a ton of acetone, I would be able to multiply that by any factor that you choose. Once the bacteriology of the process is established, it is only a question of brewing. I must get hold of a brewing engineer from one of the big distilleries, and we will set about the preliminary task. I shall naturally need the support of the government to obtain the people, the equipment, the emplacements and the rest of it. I myself can't even determine what will be required.'

Thus it was that Weizmann was given a free hand by Winston Churchill and the Admiralty and he set out by building a pilot plant at the Nicholson gin factory in Bromley-by-Bow at the beginning of two years of intensive effort. Six or seven months later, they were regularly producing acetone on a half-ton scale. It was at this time that the Admiralty decided to adapt distilleries for large scale production and also to build an acetone factory 'in one of its arsenals at Holton Heath, near Poole, Dorsetshire'.

The acetone factory was built in 1917 and was the largest building on the site. The acetone production unit was made up of eight enormous vessels, each 36 feet in diameter, together with 'cookers'

Left: A general view of the acetone factory which also shows to the rear, the cooling towers for the power station at left and a gasholder to the right.

for the grain and a huge granary measuring some 85 feet high, and 130 feet long. Initially production was based on the use of maize, all of which was imported from America, but a shortage of this material developed as the U-boat campaign reached its peak. Artichokes were used successfully as an alternative, although horse chestnuts and even acorns were also tried. John England having carried out some further research on the matter, revealed records which showed that school children were encouraged to gather acorns from the woods.

Close by, but separate from the acetone factory, was a unit concerned with acetone recovery. This unit took in fumes from the drying stoves where a good deal of the solvent was given off as vapour. This recovery plant although not as large as the production factory, contained a complex array of pumps and stills which indicated the importance which the Admiralty ascribed to the minimisation of waste and production costs.

Ultimately, when America had joined the war, acetone production was largely transferred to Canada and America. Some 20,000 tons of acetone were produced by Weizmann's process during the First World War alone and production at Holton Heath continued into the 1930s. The need for the solvent disappeared, as far as RNCF was concerned, in 1927 when a new process for the production of solventless cordite (SC cordite) was introduced.

During the war period, Dr Weizmann had many opportunities to meet leading British politicians, including Balfour – who succeeded Churchill at the Admiralty – and Lloyd-George, who became Minister of Munitions.

Upon the formation of the modern state of Israel, Dr Chaim Weizmann became its first President. His academic pursuits led him to become Director of the David Seiff Institute at Rehovat, which was later renamed in his honour. Interestingly, although the second and third Presidents were primarily politicians, the fourth President, Ephraim Katzir, was also a biochemist of international repute.

At Holton Heath, the work started by Weizmann continued and in 1922 Dr Henry 'Bill' Bunker (later Sir Henry Bunker) joined RNCF where he collaborated with Dr A C Thaysen, who was in charge of acetone production, on the microbiology of cellulosic and other textile fibres. The Microbiology Unit Laboratory associated with the acetone factory was widely known throughout the RNCF as the 'buggery' until, in 1933, it closed and work was transferred to the Chemical Research Laboratory at Teddington.

Left: This photograph shows two gas holders (No.s 2 & 3) in the foreground with a glimpse of one acetone fermentation vessel at the rear. The building seen between the two holders is not shown on the site plan of the Second World War era and the gas works was located some distance away by the main coal yard and boiler house.

Above: The acetone plant from the south, clearly illustrating the height of the building. This was the only structure on site to include a lift.
Below: This view is taken from the same area as that opposite, however the camera has panned to the right, showing both the gas holders and fermentation vessels. The dates of photography clearly differ though, since the four towers visible in the opposite view are not present.

Above: A band conveyor in the acetone factory.

Below: An interior view of the acetone plant, showing steam-heated cookers.

Acetone production

Above: Acetone factory – fermentation tank.

Above: Acetone factory – auxiliary fermentation tanks.

Above left: A fermentation tank with pipelines.

Left: Distillation columns in the acetone factory.

Centre: A flash regulator.

Right: Final acetone stills.

S2.

S3

Above: Water-spray cooling pond serving the acetone recovery plant.

Acetone recovery

Opposite above: Building S2 – an acetone recovery stove.

Opposite below: Detail showing external pipework associated with acetone recovery stove S3.

Right: Acetone recovery primary stills.

Above: Fractionating columns in the acetone recovery area.
Above centre: A view of the tops of the absorption towers.

Above: A group of acetone final recovery stills.
Below centre: Kestner fans in the acetone recovery area.

Above: Sulphur burners.

Right: A view of the factory's hydrogen production plant, with a three-stage compressor visible to the right.

DANGER

Above: The final moments of the redundant acetone factory, which was built to produce acetone by distillation from such diverse sources as acorns and chestnuts, collected by local schoolchildren.

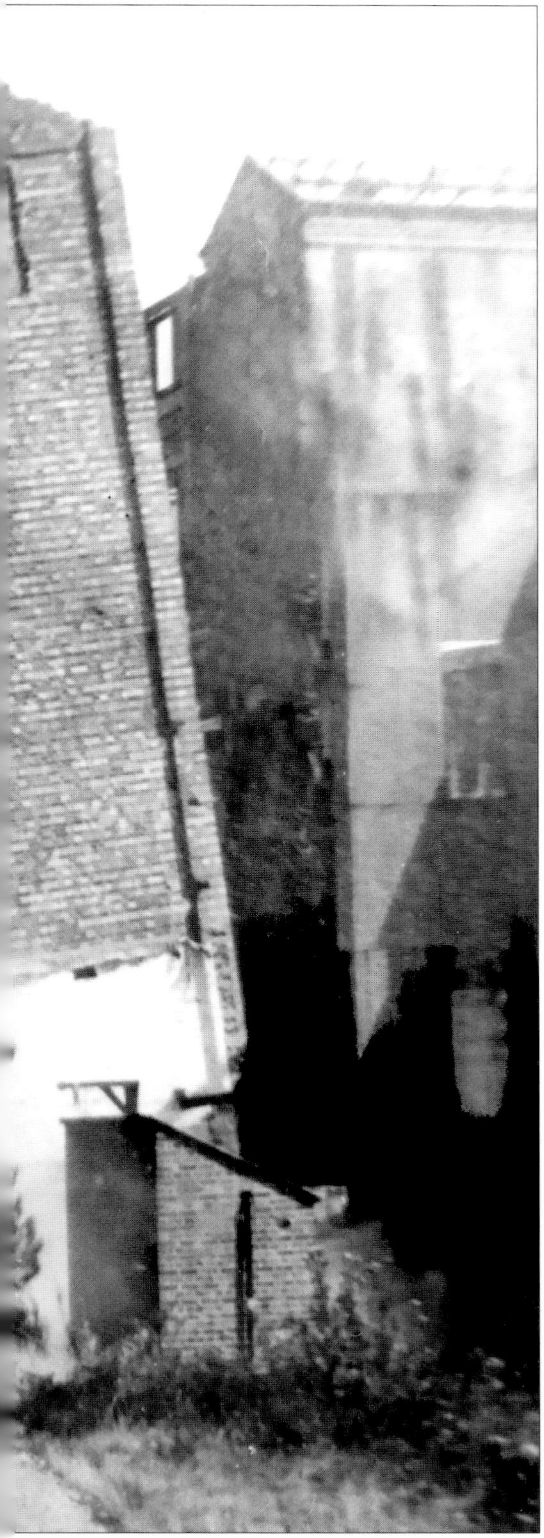

Since both the acetone factory and the laboratories associated with it were now closed, plans were made to demolish the huge granary building and at two o'clock in the afternoon of August Bank Holiday Monday in 1934. After about 18 months of preparation, the redundant structure was blown up leaving something like 5,000 tons of rubble. This rather satisfying bit of demolition work was overseen by Bill Coombes and the preparation was commenced by stripping the roof, removing windows and doors and knocking out panels. This left three main parallel walls supported by the 18-inch-thick piers alone. The piers in the rear wall were cut through and the piers in the central one had 'V' shaped cuts made in them, similar to the way that a forester prepares a tree so that it will fall in a preferred direction. The piers in the front wall were each drilled and three sticks of gelignite – each about the size of a man's forefinger – were placed in each one.

When blown, the whole building pivoted on the central wall and fell in a most satisfactory way, with a lot of noise and a great deal of dust. The huge concrete fermentation vessels remained, however, and in the Second World War four of them were suitably modified and pressed into service as air raid shelters or refuges. One was also later employed as a hydraulic testing tank. These vessels have never been removed and although now surrounded by shrubbery can still be seen.

Below: Dubbed 'The largest flower pot in Dorset', the remains of one of the acetone vessels, having been filled with soil to act as an impromptu Second World War air-raid shelter, now sports a small natural woodland.

Chapter 6

MAJOR ACCIDENTS INVOLVING EXPLOSIVES

In relation to the numbers employed at the factory and the length of time it was operational, very few people lost their lives as a direct result of explosives. However, of the fatal accidents, the most serious was the Nitroglycerine Hill disaster which occurred at 10:43am on Tuesday, 23st June 1931 when a charge of nitroglycerine – about one and a half long tons – blew up and ten Nitroglycerine Hillmen lost their lives.

The nitroglycerine workers would have realised only too clearly that sudden death was just one careless move away, but fifteen years of trouble-free production must have led to a feeling of considerable confidence, both in the plant and in the well disciplined RNCF organisation. There must also have been confidence in the people. The Manager, Mr W T Thomson, was a prominent member of a celebrated family in the explosives world and Mr Blair, the Chief Chemist was competent and dependable. The principal foreman was Ernest Norton, a man who had been with the RNCF from the beginning, when he was assistant foreman in the nitroglycerine section to a man named Hollingberry. 'Nobby' Norton was a regular churchgoer and a highly respected Nitroglycerine Hillman totally reliable and with the vast experience borne out of almost daily contact with this treacherous explosive over many years. It is true that he often told Bill Coombes that nitroglycerine would be the death of

Below: A view of the site looking from the main reservoir, towards the Baker's Arms pub. Taken in the early years of the factory, this photograph shows in the foreground the nitroglycerine plant which was to explode in 1931. The extent to which the standard-gauge railway penetrated the site is illustrated by the row of trucks in the distance.

him, but such expressions are often used without being believed by either speaker or listener.

On Tuesday, 23rd June, the day broke with warm sunshine and a council workman in Poole was decorating the High Street in preparation for the Annual Carnival to be held the next day. It seems probable that some of the Nitroglycerine Hillmen had made plans to enjoy some part of the day's entertainment after they had finished their shift. At Poole Grammar School, Bert Wareham and other boys were playing outside during the morning break.

On a farm at Lytchett Matravers, Hugh Dennet was happy about the sunshine as it was ideal for haymaking, and it would be soon finished if the weather held. At Townsend – in Corfe Castle – Geoff Groves, then a toddler of three or four years, played at home with his mother.

At the factory, something had occurred that required the presence in the AB house of both foreman Norton and Chief Chemist Blair. There was obviously something amiss with the morning shift's nitroglycerine production and the workman involved who had reported the problem had been sent on another job, away from the Nitroglycerine Hill area.

Suddenly, at 10:43am precisely, as Frank Marshalsay was passing the RNCF on his motor cycle, he found himself shaken violently off his machine and onto the A351. Looking up afterwards, he saw a lurid column of smoke rising from the factory. The haymakers at Lytchett felt the same shock and saw what appeared to be red flames – but were in fact oxides of nitrogen – rising in the wake of a massive explosion.

The workman in Poole High Street and Bert Wareham and his schoolmates had to wait a few seconds before they too heard the explosion and their eyes were drawn automatically to the direction of Holton Heath. Immediately, boys with relatives at RNCF fell to speculating and wondering. Meanwhile, the council employee in the High Street dashed into the nearby offices of the *Poole, Parkstone & East Dorset Herald* with the news. Directly across Poole Harbour, Geoff Groves recalls his mother in tears, rushing to the top of Townsend Road in Corfe, already convinced of the fact that her husband Reg – a Nitroglycerine Hillman – would not be returning.

Ted Terrell was in the main laboratory when the explosion occurred and he recalled it thus: 'Suddenly there was an unearthly silence; everyone looked at each other, then WOOF!' Being trained as fire fighters we instantly took the fire hoses and made for the nearest point to the disaster area. After running for some time Mr Hazell, Chief Chemist, who was in the lead suddenly put his hands to his head and shouted 'Go back! Go back!' It appears that after the heavy debris had fallen, the acid was coming down as a fine spray. As this was about ten minutes after the explosion it demonstrated how high the contents must have been blown. This acid subsequently set fire to the dry heath. It is probable that the 'unearthly silence' described by Mr Terrell was due to the inrush of air to the site of the explosion.

At the factory, it may be guessed that the management got as near panic as it is possible to imagine that such a well-disciplined organisation could get. All those for miles around with relatives or friends working at RNCF would have heard the explosion, felt it and seen the result of it in the sky. Fear and anxiety would have been dominant emotions for these people at that time and that was how it was to be for several hours. The Royal Marine Police, who at that time were responsible for the security of the factory, locked all the gates.

More detailed news of the disaster leaked out through the black, 12 feet high fence surrounding the factory, as survivors arrived at the surgery near the main gate. An official list of the dead and missing was ultimately made available just before tea time. The delay arose out of the fact that such a list could only be prepared after a roll call, since the violence of the explosion precluded the possibility of counting corpses. The list was horrifyingly long with ten names upon it. Norton and Blair were on the list, of course, but sadly Walt Brown, (assistant Foreman) 'Jock' Henderson, Fred Lowman, Bill Vivian, Bert Guy, Reg Groves, Bob Taylor and Bill Hickman – all Nitroglycerine Hillmen – were also included.

It is often the case that, because of an accident or some other unplanned event, individuals are spared a particular tragedy but it cannot be often that life or death has been unwittingly decided by the spin of a coin. However, this is what happened when the work was allocated by Walt Brown in the changing room at the start of this fateful shift. Harry Gover and Bill Hickman were instructed to cover filter house 'C4' and perform other jobs around the section generally. Gover and Hickman tossed a coin to decide who did what and Hickman went to C4 and was killed and Gover survived to tell the tale.

A total of 19 were injured, but curiously the most severe case was a broken leg, sustained by Arthur Arkell who was struck by a chunk of lead while working in an area comparatively remote from the seat of the explosion. A 'Chippy' – Bill Cave – was unharmed, except that his shirt was taken neatly off his back. Many others had holes appear in their clothes as acid, from tanks ruptured in the explosion, rained over the factory area.

A tramp was passing the perimeter fence at the time of the explosion, about half a mile away. Shortly afterwards, he hammered

Above: The remains of mixing house 'E3' after the 1931 explosion. Although protected from the blast by distance and earth embankments, the nitroglycerine contaminated guttering was responsible for conveying the explosion to this and other areas. The view at rear shows the reservoir hill.

Above: The remains of the charge house which supplied the AB house, after the 1931 explosion. This picture was taken from the edge of the crater where the AB house once stood. The wooden charge house building has vanished, leaving the distorted tanks it once enclosed.

on the locked gates shouting 'Someone's shot one leg of my trousers off!' He was told 'Come back tomorrow, we can't deal with you now.' However, the tramp turned up at the office still with one bare leg, demanding 'What are you going to do about my trousers?' He was taken care of by one of the police officers and installed in a disused nissen hut. A 'whip round' by the officer and main laboratory staff provided him with a completely new rig out, including boots. He was given odd jobs to do, but when he was told he was to be given a permanent job, he disappeared as quickly as he came!

It started raining the next day and it appears that no-one was very enthusiastic about the carnival. The inquest was held at RNCF and the funerals were held in the presence of large congregations. Local papers published long and – by today's standards – mawkish articles. The *Daily Sketch* arranged for an aeroplane to fly over the site and secured an aerial view of the devastation which they published. This was an illegal act and the newspaper was subsequently prosecuted and fined £500. The occurrence of the explosion was also recorded in *Whitaker's Almanac* for 1932.

According to the unofficial notes of Dr Ramsay – a nitroglycerine chemist at RNCF in the late 1930s – the following appear to be the general facts surrounding the disaster. One charge of nitroglycerine had already been made and was standing in the nitrator separator awaiting pre-washing. The worker arrived to prepare the second nitrator separator for a charge and to do this the waste acid, which remains in the nitrator after nitroglycerine has been removed, had to be run off and replaced with fresh acid from the charge house. When the time came to carry out this operation he went below, opened the cocks and ran off the contents to the 'egg' (technical jargon for the container designed to manipulate nitrating acids). On returning to the top platform, the worker found that his nitrator was still full of waste acid, but that the other was empty. This meant that he had run waste acid mixed with nitroglycerine into the acid 'egg'.

The method used for getting the contents of the 'egg' back into the nitrator involved transferring the liquids via the charge house, under the influence of compressed air. The intention was to blow the contents to the waste acid displacement tank and at the same time to open the displacement cock so that the contents would run down to the nitrator. Blowing acid from the 'egg' to the charge house requires manipulation of cocks at each end, with a man part way along the gantry to signal. This man was visible to another worker in washing house 'C2' who thus knew what was going on, when it commenced and how long the operation had been in progress.

It had been in progress for about 10 minutes when the explosion occurred and the AB house, Charge house, Filter house C4 (near the glycerine and soda store) and Mixing house E3 were all destroyed. The Filter house and the Mixing house were involved because a portion of nitroglycerine had been sent from the AB house to the Filter house and another portion had been sent from the Filter house to the Mixing house and in neither case had the gutter yet been washed down. The fact that the gutters propagated the explosion was demonstrated by the destruction of the canvas flap along its whole length, by the temperature

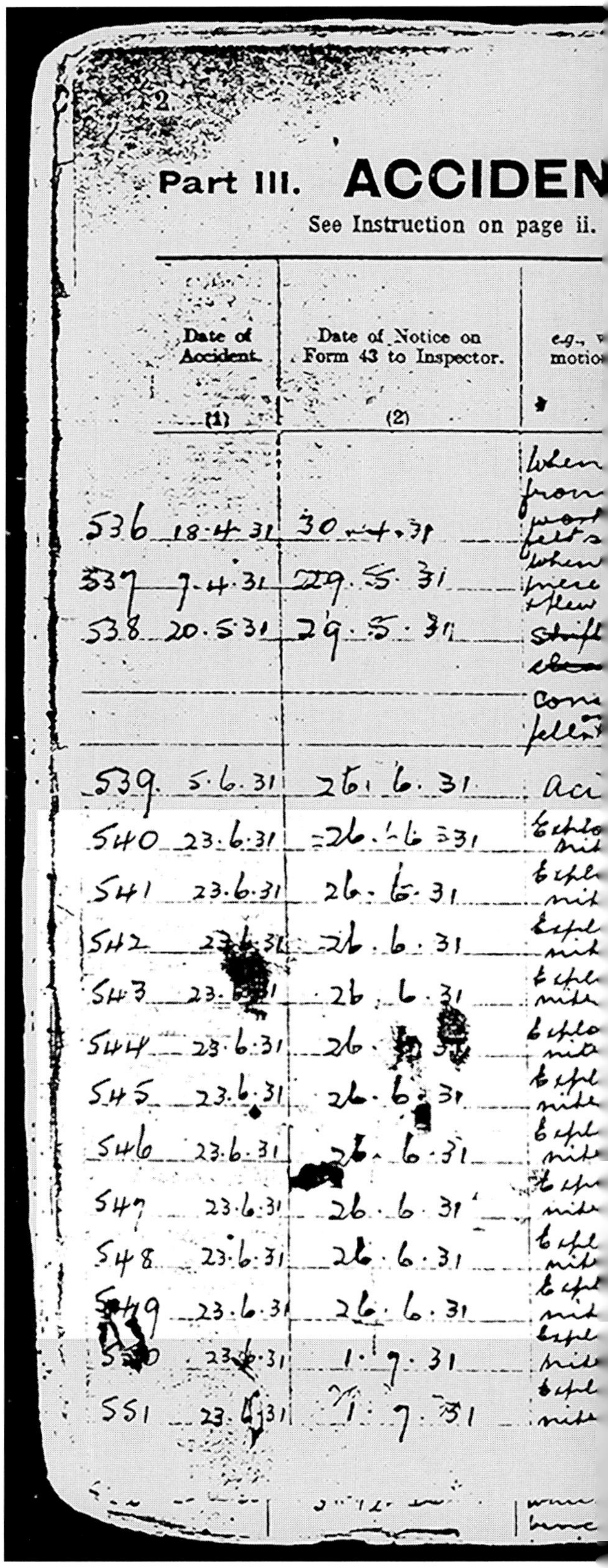
Part III. ACCIDEN

See Instruction on page ii.

	Date of Accident. (1)	Date of Notice on Form 43 to Inspector. (2)
536	18.4.31	30.4.31
537	7.4.31	29.5.31
538	20.5.31	29.5.31
539	5.6.31	26.6.31
540	23.6.31	26.6.31
541	23.6.31	26.6.31
542	23.6.31	26.6.31
543	23.6.31	26.6.31
544	23.6.31	26.6.31
545	23.6.31	26.6.31
546	23.6.31	26.6.31
547	23.6.31	26.6.31
548	23.6.31	26.6.31
549	23.6.31	26.6.31
550	23.6.31	1.7.31
551	23.6.31	1.7.31

Right: An extract from the factory accident records which clearly details the effects of the explosion upon the workers.

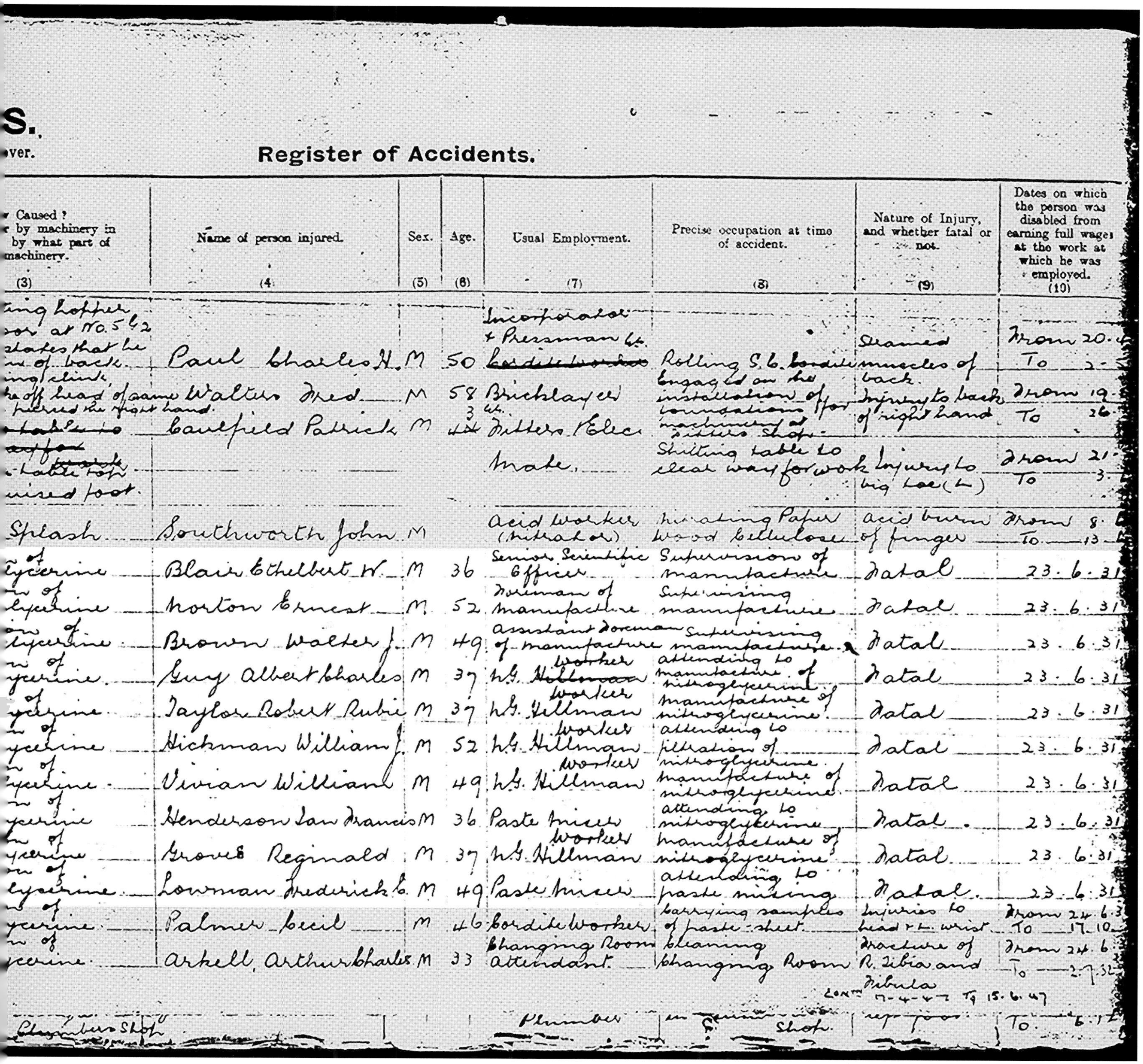

S.

Register of Accidents.

Name of person injured. (4)	Sex. (5)	Age. (6)	Usual Employment. (7)	Precise occupation at time of accident. (8)	Nature of Injury, and whether fatal or not. (9)	Dates on which the person was disabled from earning full wages at the work at which he was employed. (10)
Paul Charles H.	M	50	Incorporated & Pressman &c. ~~Cordite Worker~~	Rolling S.C. ~~Cordite~~	Strained muscles of back.	From 20. To 2.
Walters Fred.	M	58	Bricklayer &c.	Engaged on the installation of foundations for machinery at Fitters Shop.	Injury to back of right hand	From 19. To 26.
Caulfield Patrick	M	43	Fitters & Elec. Mate.	Shifting table to clear way for work	Injury to big toe (L)	From 21. To 3.
Southworth John	M		Acid Worker (Nitrator)	Nitrating Paper Wood Cellulose	Acid burn of finger	From 8. To 13.
Blair Ethelbert W.	M	36	Senior Scientific Officer	Supervision of manufacture	Fatal	23.6.31
Norton Ernest	M	52	Foreman of manufacture	Supervising manufacture	Fatal	23.6.31
Brown Walter J.	M	49	Assistant Foreman of manufacture	Supervising manufacture	Fatal	23.6.31
Guy Albert Charles	M	37	N.G. ~~Hillman~~ Worker	attending to manufacture of nitroglycerine	Fatal	23.6.31
Taylor Robert Rubie	M	37	N.G. Hillman Worker	manufacture of nitroglycerine	Fatal	23.6.31
Hickman William J.	M	52	N.G. Hillman Worker	attending to filtration of nitroglycerine	Fatal	23.6.31
Vivian William	M	49	N.G. Hillman Worker	manufacture of nitroglycerine	Fatal	23.6.31
Henderson Ian Francis	M	36	Paste Mixer	attending to nitroglycerine	Fatal	23.6.31
Groves Reginald	M	37	N.G. Hillman Worker	manufacture of nitroglycerine	Fatal	23.6.31
Lowman Frederick C.	M	49	Paste Mixer	attending to paste mixing	Fatal	23.6.31
Palmer Cecil	M	46	Cordite Worker	carrying samples of paste sheet	Injuries to head & L. wrist	From 24.6.3 To 17.10.
Arkell, Arthur Charles	M	33	Changing Room attendant	Cleaning Changing Room	Fracture of R. Tibia and Fibula	From 24.6. To 2.7.32
			Plumber	Shop		To 6.1

TEN DEAD IN NAVAL FACTORY EXPLOSION

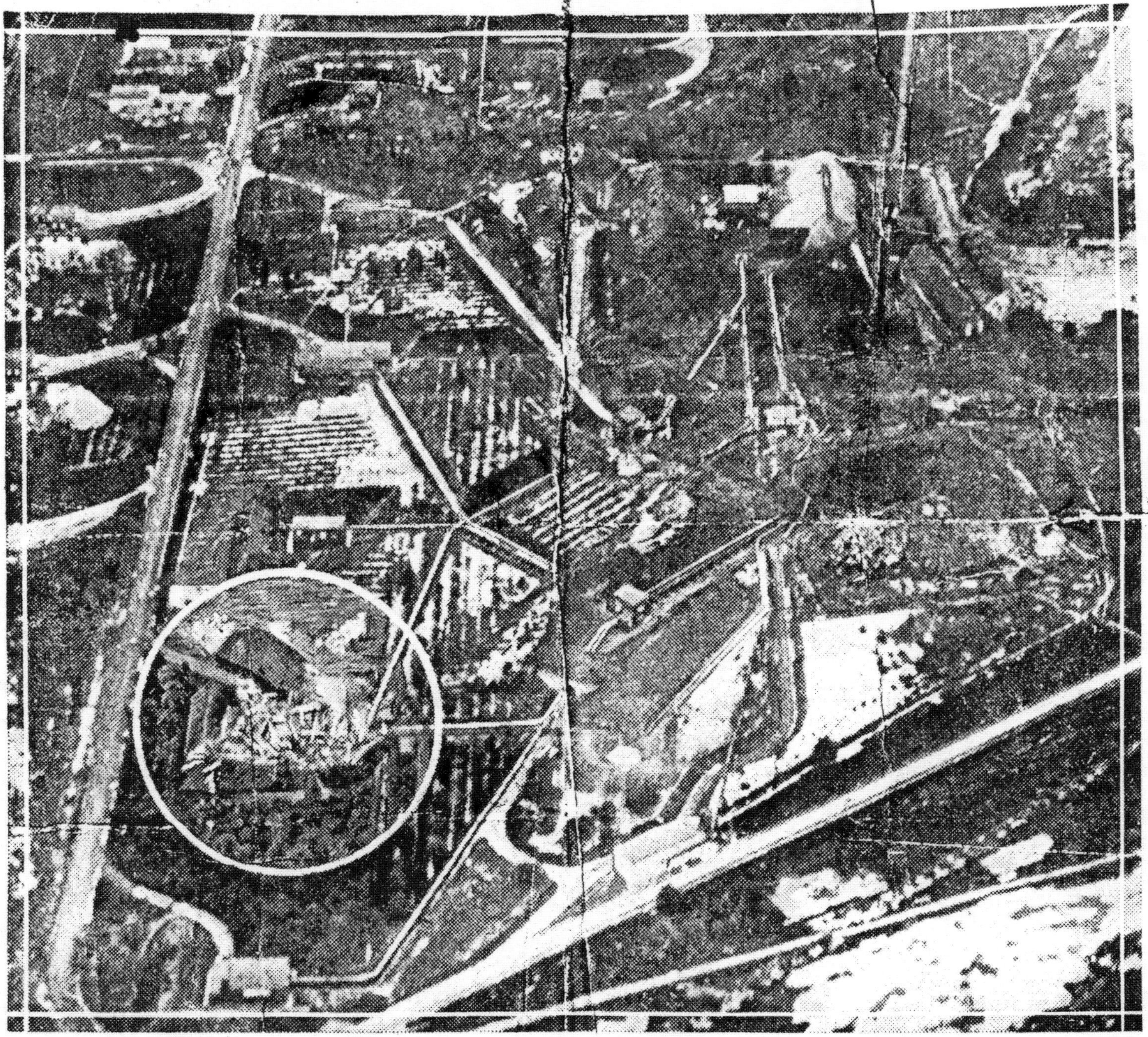

Ten men were killed and many injured in an explosion at the Naval Cordite Factory at Holton Heath yesterday. This *Daily Sketch* aerial photograph shows how the effects of the explosion (marked with a circle) were localised by the protective earthworks, which are a feature of the lay-out of the factory. The force of the explosion was felt five miles away.

June 23rd 1931 10 Killed 19 Injured.!

Opposite: A reproduction of the *Daily Sketch* photograph, which wrongly identifies mixing house E3 as the seat of the explosion. The crater left by the AB house is discernible at centre right.

Below: A view of one of the batch-wise nitroglycerine plants. This is believed to be the AB house which was involved in the 1931 explosion. The two handwheels of the drowning cocks are visible in the centre. The brine control valves are at the ends of the platform. The sight glasses and short gutters along which the nitroglycerine runs into the pre-wash tank may also be seen. Note the large earthenware stop-cocks, used to control the entry of mixed acid to the nitrator vessels.

of the gutter afterwards and by depressions in the gutter by each transverse butt weld where a little nitroglycerine lodged.

It seems likely that the explosion was initiated when the egg was just emptied and a rush of air forced the last of the liquid through. This would have been nitroglycerine and it would have slammed against the mild steel waste acid displacement tank with great force. There was the further possibility that it could have been caused by friction when a stopcock was manipulated at the end of the transference operation.

With hindsight, it seems a rather risky business to force

nitroglycerine around the place with compressed air and, as the top of the egg could be removed, the contents might have been more directly and safely removed. Percy Bartlett – a plumber who worked for 30 years on the site – is of the opinion that the former technique had been used with success in the past and having 'got away with it' once, the method was used again. It is worth noting that Reg Groves was intending to be transferred away from the Nitroglycerine Hill, as he considered a certain member of the team to be what these days would be called a bit of a 'cowboy'.

The force of the explosion was well illustrated by an exhibit which was for some time on display in the RNCFA museum. This was a heavy brass gong which had been penetrated right through by a fragment of softwood, which still remained in the hole.

As a result of this tragic accident, a number of alterations were carried out in the remaining nitrator separator house to make the process safer. For example, the two nitrators were separated above and below so that a worker would have to make a deliberate detour to get from one to the other. All cocks on the nitrators and those in the charge house were fitted with a locking device and the keys were shaped so as not to be interchangeable. These keys were carried on a lanyard worn around the neck of the appropriate workers.

A third modification involved the removal of all cocks between the 'egg' and the displacement tank. There were also other, technically rather obscure changes, but those already mentioned were most important. A procedural change after the explosion meant that a chemist was always to be present during the nitration process. In addition, in order to prevent explosions spreading to adjacent buildings, the mechanical transport scheme mentioned earlier was introduced to carry nitroglycerine from wash houses to mixing houses.

The report of the Committee of Enquiry revealed that by mistake, a mixture of nitroglycerine and acid was allowed to run into a trap not intended to receive it and that the operation of forcing the mixture back into the tank from which it came caused the explosion. When the inquest was resumed at RNCF on Thursday, 13th August, the coroner and the jury debated for an hour the wording of the verdict which the latter returned. The jury wanted to attribute the cause of the explosion to the original mistake. The jury retired four times and finally, the coroner accepted their verdict that death was accidental and due to an explosion caused by an error in the process of manufacturing nitroglycerine. They added a rider that no blame could be attached to any of the deceased men.

The 1917-8 accident

Although this was by far the most serious accident to occur at the RNCF, there were a number of others which involved fatalities and the first of these happened in late 1917 or early 1918 when Jesse Orchard from Poole was killed in an explosion. Jesse was a temperature taker, which meant that he had to proceed from stove to stove on his rounds, checking the temperature at which the cordite was being dried. When the drying stoves had been loaded with cordite – fresh from the acetone recovery process – they were then sealed and the final drying commenced. This stage could take several weeks, but the precise time depended upon the diameter of the cords. The drying was accomplished by the use of steam-heated pipes which passed through the buildings. The temperatures were taken without entering the stove but, for some reason or other on this particular night, the contents of drying stove No.17 exploded just at that time when Mr Orchard was in attendance.

The thud of the explosion woke Bill Groves in Upton, who got to the bedroom window in time to witness what looked like yet another spectacular firework display.

The 1927 accident

The factory then enjoyed a disaster-free period of about 10 years until 1927, when the RNCF was changing over to the production of solventless cordite and the acetone recovery house and associated plant had become redundant and was to be dismantled. By the morning of Saturday, 10th December, this work was well under way. Victor Coward – a plumber from Lytchett Minster – and his mate, Walter Barnes from Poole, had already finished working at No.4 stove and by 11:00am were seen starting to dismantle the acetone vapour pipework between drying stove No.5 and the acetone recovery house.

The acetone vapour pipes were supported at intervals by stanchions at a level of about 8 feet from the ground. Stove No.5 was full of cordite and workers were busy within it when the tradesmen arrived in the company of Ernest Lovell, a 50-year-old, highly experienced cordite worker from Wimborne. It was a rule of the factory that tradesmen should always be accompanied when in the Danger Area by suitably qualified explosives workers. One of Mr Lovell's tasks was to regularly inspect the traps which were included in the acetone vapour pipe work and to collect any nitroglycerine that had condensed within them.

Right: The 'new' washing house C2 which was built following the explosion. Although the building is long gone, the tunnel to the right of the picture has survived.

Left: The memorial service for those who died in the 1931 explosion, held in front of the administration building at Holton Heath. The band is believed to be from the Royal Tank Corps of Bovington.

For the more technically minded it should be explained that, although in the early stages of drying the cordite almost pure acetone vapour was given off, small quantities of nitroglycerine vapour were also drawn off the product, especially towards the end of the drying operation. This nitroglycerine vapour was carried into the acetone vapour recovery pipes where it condensed and hence Mr Lovell's daily round with a rubber bucket and a very steady hand. However, his function on this particular Saturday was to ensure that all went well and that the plumbers did nothing that was unduly risky.

Just after the tradesmen had started dismantling the pipework, three women workers left stove No.5 to clock off a little earlier than usual and they noticed the men singing at their work. At about this time, Tom Prince and Sidney Brown went into this stove.

An explosion occurred at about 11:10am and this was immediately followed by the roar of huge flames as the cordite in the drying stove was ignited. Flames were seen leaping into the air to an estimated height of 100 feet. Messrs Prince, Brown and another worker, Davies, were blown right out of the building but thankfully were not badly hurt. Not surprisingly, however, the three working on the acetone vapour pipes were killed instantly by the explosion.

Mr Spear of Wareham – who was a factory messenger at the time – was responsible for setting up the telephone call between Holton Heath and the Admiralty in London when the time came to report the accident. He was present when Commander Priston spoke to their Lordships and clearly remembers the opening phrase:

'I regret to have to inform you of an explosion and death...'

The Manager and Deputy Superintendent of the factory, Mr W T Thomson, later estimated that about half a pound of nitroglycerine would have been involved. He also suggested that perhaps the disaster may have been related to slight subsidence of the stanchions carrying the pipes during the recent heavy rains. As a result of this, nitroglycerine would have collected at a low point and could well have seeped into a pipe joint. Friction resulting from unscrewing the joint would have been quite sufficient to detonate the explosive.

Mr Thomson went on to say that there had been no urgency associated with the work and that if regulations were closely adhered to then the work was fairly safe but things would go wrong sometimes without the cause being known.

The chemist in charge of the cordite section at this time was a Mr H H Hazell and it is said that he personally completed the similar work on the remaining stoves. The footings of stove No.5 were used later to house a small laboratory and renamed 'X1'. This was a small wooden building with a large 'witches hat' type ventilator on the top. This was popularly known as 'Thomson's Tabernacle' owing to its church-like appearance. The earth traverses surrounding this building were well known as a refuge for adders during the last war and still are today. Nothing now remains apart from a pile of rotting wood, the old ventilator and, of course, the adders.

The transition to the solventless cordite (SC) process proved in the end to be a very painful one, for just over two years later, on Thursday, 7th November 1929, an explosion occurred in press house No.8 and as a result another man lost his life and three others, including two women, were injured.

The new SC process involved the mixing of nitroglycerine with a slurry of nitrocellulose in water, followed by stabilisation with Carbamite and a number of other operations dealt with elsewhere in this account. The end result was a 'cheese' of cordite, made up in layers, which was then taken to a press house by an electric locomotive and kept warm by means of hot water jackets. At the press house, the 'cheese' was kept in a warm oven until placed in the cylinder of a press.

According to the second pressman, Thomas Littlechild, who had just finished pressing cordite on press 'B', George Battrick was commencing work on press 'A' when there was a flash and a terrific explosion accompanied by a partial collapse of the building. Mr Littlechild followed Miss Sheriff who was also working on press 'B' through the debris and out of the building. Miss Sheriff would have been assisting the pressman by cutting the long strips or 'cords' to length and packing them into boxes.

Joseph Turner, who was a fitter at RNCF, was working in Press House No.11 at the time and he immediately ran to the scene of the accident. On entering the house, he saw Mr Battrick sitting on the floor with burning debris on top of him. According to Mr Turner, who gave this evidence at the inquest, he then dragged him clear and tore off some burning clothing, after which Mr Battrick was taken to the factory hospital. It is sad that this courageous act – for there was still cordite within the building – did not have a happier ending but the fact is that George Battrick, who was transferred to the Cornelia Hospital in Poole on Friday, died at 9:40pm on Saturday, 9th November. Mr Turner was subsequently presented with the Edward Medal, later to become the George Medal, for this act of bravery.

Miss Norah Bishop was also assisting in Press House No.8 on that day, presumably on press 'A', and of the survivors, she was the most seriously injured. She was very severely burned and was kept in the factory hospital for several weeks during which time she lay on an experimental water bed devised by Dr Thaysen, a microbiologist at the factory. It is also understood that she was the subject of some early skin grafting. In any event, despite her very serious injuries, she did recover and later married another employee at the RNCF.

The cause of the accident was, according to the *Poole & Dorset Herald*, partially explained at the inquest. Mr Thomson – the MDS at the factory – said that the operator had done nothing to cause the explosion and that the press was at fault. He said that experiments had been conducted in which cordite had been safely pressed at twice the cut-out pressure of 900psi.

An official enquiry was undertaken into the cause of the explosion although its findings were not made public. Discussion with old cordite workers has suggested that a nitroglycerine-rich composition may have been to blame. It is understood that the cordite batch was deficient in nitrocellulose by one bag. In those circumstances the cordite would have been a much more hazardous material. In fact after this incident a QE (Quick Extract) test was introduced to determine the nitrocellulose content of each batch of cordite. A card bearing details of the nitrocellulose content travelled with the batch as it passed from stage to stage in the manufacturing process.

There were a number of other modifications as a result of the tragedy, most of which involved refinements in press house design. For example, press houses were subsequently arranged so that pressing, cutting and packing operations were all kept separate, although still within the same building. The other change involved a re-design of women's working clothes. Until this time, women cordite workers wore blue serge uniforms – hats, coats and skirts – but, because the extreme severity of Miss Bishop's injuries was at least partially attributable to wearing a skirt, trousers became a regular part of women workers' uniforms from then on.

It is of interest to note that process workers in hazardous industries, although being in the 'front line' should an accident occur, are generally the lowest paid people on site. An example is provided by our more modern nuclear industry, where process workers, one slip away from extremely toxic plutonium, get only slightly more pay than the general worker, on the lowest pay scale of all.

Plumbers, perhaps understandably, did seem to be especially vulnerable in the factory and in 1934, on the afternoon of Monday, 23rd July, another was killed while working in the lead pound. On this occasion, the plumber – Francis 'Jack' Mitchell from Upton – was sawing through a lead pipe which had been used to circulate air

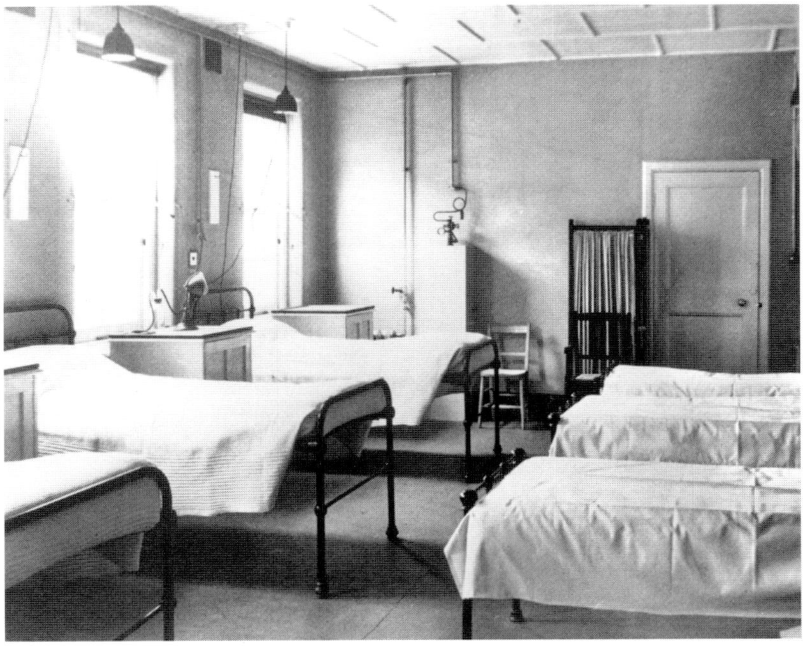

Above: Part of the factory hospital. Well equipped for its time, it was used for some pioneering work on skin grafting following a serious accident involving a cordite fire. Note the unusual ceiling which was a mixture of plaster and asbestos. Such ceilings were common in all of the non-industrial rooms at RNCF – a feature also employed within the RNCF cottages.

in a pre-washing tank, used early in the nitroglycerine manufacturing process. It was concluded that a bead of nitroglycerine must have remained, lodged somewhere within the pipe, in spite of earlier attempts to render any traces of explosive harmless by washing with soda in the approved manner. The accident was particularly sad because it was only a minor explosion and had Mr Mitchell been even a little farther from the pipe he would probably have survived. Mr Mitchell's workmate, Frederick Barnes, who was also in the pound at the time, sustained only slight injuries to his forearm and was able to go home at the end of the shift.

After this accident, all similar operations had to be carried out in the presence of a chemist and a special piece of equipment was erected in the lead pound which enabled the plumber to saw from behind a barrier, while a water jet played upon the work piece. Another safety measure involved the application of a 'long flame' – a blowtorch on a sufficiently long pole – in order to set off any remaining material from a safe distance.

Thankfully this was the last fatal accident to occur at the RNCF though this is perhaps surprising in view of the high production rate sustained over the long and stressful Second World War.

Chapter 7

THE NEW NITROGLYCERINE HILL

For a time after the 1931 disaster, the factory had to operate within the limitations of the nitroglycerine output from the sole remaining plant in the ABC house (renamed 'AB1' on latest site plan). Meanwhile, plans were being made for the building of a new plant on the site of the old AB house, to be designated 'AB2' but, before this could be done, the site had to be cleared since all that was left intact was the concrete lined drowning tank.

It fell to Bill Coombes and others from the Works Department to carry out this preliminary work, but in order to enter a danger building of which the drowning tank was once a part, it was necessary to get a clearance certificate. This was initially refused by Mr Whittaker, who had taken over from the late Mr Blair, on the reasonable grounds that no one had examined the spot since the explosion and therefore it could not be certified as clear. However, it was believed to be safe and in order to make sure, a ton of caustic soda was used to ensure

Below: The new Nitroglycerine Hill. A remarkable picture which shows the vast civil engineering exercise which went into the building of AB2, the Schmid nitrator building. This view is taken from the main reservoir hill, looking towards Lytchett Minster. The hill remains, as does the white glazed underground building. Now tree covered, the hill appears as a natural feature.

Above: A progress photograph, showing picrite compressor house N1 nearing completion. Although the track bed remains, every trace of the building has been removed. Note the cooling pond (heat exchanger) for the refrigeration compressors housed upon the roof.

'neutralisation' of any remaining traces of explosive. Nevertheless, works staff were still reluctant to go down into the tank, even though Mr Whittaker volunteered to stand at the top! Anyway, the matter was ultimately resolved, clearance was obtained and the rubble was removed and much old concrete was knocked out without further mishap.

Meanwhile, it had been agreed that the new plant would be of the Schmid type which yielded a continuous supply of nitroglycerine as opposed to the older Nathan-Thomson-Rintoul plant which embodied a batch-wise production method. Once this decision had been made, the Works Department prepared drawings for a proposed new nitrator-separator house to be built on the site of the old installation in order that it might fit in with the existing layout. The design was for a building, essentially similar to the old AB house, in the sense that it was to be constructed of similar materials, but it would be built with the lower half below ground and the upper half above. The whole was to be encased in a further containment of reinforced concrete finished within by glazed white bricks. Entrance was by two similarly constructed and finished tunnels at right angles.

The result was a wooden two-storey building constructed inside a concrete building, leaving a gap between the two of approximately 3-4 feet. The entrance tunnels led into the upper-storey and steps led down to the cellar, which contained the original drowning tank, retained from the previously destroyed plant.

The 12-inch-thick concrete over the crown of the building was reinforced with two rows, at right angles, of half-inch diameter steel bars at 6-inch centres. The walls or the concrete shell were much thicker, ranging from 21-inches at below ground level to not less than two feet thick above ground level. Earth protection to a minimum of 40 feet was provided over the building, partly to offer a measure of protection against aerial attack and lightning strikes and partly to localise the effects of any explosion within the building.

The restricted area available for the spread of this earth cover called for a much steeper slope, 1:1.25, than the natural angle of repose and in order to overcome this difficulty, frequent layers of brushwood were placed at right angles to the face of the slope. To build the earthwork, soil was brought from a remote and unused part of the factory in skips, hauled by small Ruston & Procter diesel-powered locomotives on a narrow-gauge (2 feet 6 inches) rail loop laid especially for the purpose. As the hill grew higher, the railway had to be arranged in a spiral fashion in order that the locomotives could draw the loaded skips – each of which held a cubic yard – to the top. Thirty-five thousand cubic yards of soil were used in all to complete the work. The complete undertaking cost around £9,000.

This most interesting building exists to this day and is to be seen on the right-hand side when going along Blackhill Road in the direction of the railway. It is in fact just behind the building which once housed the British Gas offices and today looks like a large, natural hill covered with trees. When originally erected, it was seen as a heather-covered hill that tended to draw the eye from any part of the factory.

When the plant was decommissioned, the lead which covered the entire floor was removed by Percy Bartlett. The internal wooden building was removed and the glazed containment filled with soil to the level of the entrance tunnels. Explosives were then detonated in the roofs of the two approach tunnels, to seal the entrances. This precaution was clearly desirable at the time, as there may well have been a fair quantity of spilt explosive remaining, especially in the lower level around the drowning tank, also the 20-30 feet drop at the end of the entrance tunnel in virtual darkness, could easily have killed a curious trespasser.

Due to the great strength of the building, the roof of the southern tunnel did not completely fall and those people who do not suffer from claustrophobia can, on obtaining permission from the landowner, still effect an entrance, although there is little to be seen now. It is quite possible that this building may remain for some time to come, due to the engineering difficulty of breaking it up. A further protection was added in the late 1990s when the structure was scheduled as an ancient monument! In addition at this time, some other and rather finer buildings which remained were also listed.

An additional safety feature of the new AB2 was the fitting of a 9 inch water main leading directly from the adjacent reservoir. This could be used to flood the building to a depth of about 10 feet in an emergency by operation of a cock outside the hill. This fact got to be known outside the factory and as it was passed on from one person to another, so it became exaggerated. Eventually, it became the custom for coaches on evening mystery tours to pause on the A351 outside the perimeter fence and for the driver to tell the tourists about the factory and how it could be entirely flooded in a few minutes should the need arise.

The new continuous nitration Schmid plant – named after its inventor – was then ordered from the Meissner organisation of Cologne and three Germans came over to oversee its installation. Herr Reber was an engineer and his colleagues were Herr Bresser, a

Above: The interior of one of the compressor houses. These units supplied the copious amounts of compressed air used for washing nitroglycerine and for the transport of liquids. Ammonia compressors were also employed for cooling brine.

chemist, and Herr Wieselhaus an expert in welding stainless steel. The work commenced during the latter part of 1935 and was completed by mid-1936. These three men were to be accompanied wherever they went within the factory by Roy Bartlett, who worked at RNCF for many years. This precaution was, of course, in the interests of security, but judging by the reaction that their presence provoked it could be equally argued that it was necessary to ensure their safety. By 1936 Nazi Germany was very unpopular and many of the workers at RNCF had fought in the 1914-18 war and had long memories. As the Germans walked around the factory they experienced a hostile reaction wherever they went. Scowls and insults were commonplace and occasionally workers spat at their feet.

However, the Schmid plant was finally installed and the first production run of nitroglycerine commenced on Tuesday, 16th February 1937. This unit was favoured from then on and bore the brunt of production all the way through the Second World War although the older ABI (formerly ABC) house continued to do sterling work. In particular AB1 was extensively used during the period March 1942 to late 1944 and less frequently into 1945. It was necessary to change the Schmid nitrator in early 1942 and so AB1 resumed manufacture and for a few weeks carried the entire production load.

For a considerable time, two charges a day (about three tons) were run at AB1 and the nitroglycerine was sent to C3 to feed El which made cordite HSCKIT for use in the eight-barrelled pom-pom gun. Because AB1 was open to the sky and not protected by earthworks to the same extent as AB2, it was the practice to ring the local anti-aircraft Gunnery Operations Room (GOR) to check for enemy aerial activity in the vicinity before commencing nitration. Production of nitroglycerine from the Schmid plant in the AB2 house ceased on Wednesday, 3rd April 1946.

Old records show that the staff at the RNCF were most interested to learn of an explosion concerning a Schmid plant that had been installed by Meissner at Nobeoka in Japan to see if any lessons could be learned. Herr Meissner's report to the Admiralty makes interesting reading and the reader is left in no doubt of his opinion of the Director in charge of the Japanese plant. To sum up the report, it would seem that there was reason to believe that the glycerine supply may have contained water but despite this, plus the fact that daylight was fast disappearing (there being no artificial light available), the Director insisted upon a nitration being carried out. Red fumes from the nitrator were visible at one stage, a sign of dangerous decomposition of the material, and in the bad light an error of judgement regarding temperature seems to have been made which caused the Director to insist that the coolant be shut off and the contents of the nitrator be immediately transferred to the separator which was already over-full with nitroglycerine from previous nitrations. As a result, the reaction became unstable and the plant exploded. Fortunately, the staff just had enough time to escape. Although the notes of the Manager and Deputy Superintendent of the RNCF shows that he was not entirely convinced by Herr Meissner's blanket condemnation of the Japanese Director, the clear fact emerged that the Schmid plant was not responsible for the incident, doubtless to the relief of RNCF staff. Of particular interest was the fact that the nitroglycerine which was well mixed with water in the washing columns did not explode.

Of AB1 there are only the tree-covered earthworks to be seen. The wooden building has gone and the area inside the earth surrounds is filled with soil. Trees have grown on this to such an extent that it is quite difficult to locate. The brick entrance tunnels survive and most surprisingly, the charge house stood until the summer of 1987 when the rotting timbers finally gave way. This building was made entirely of wood and must be of very early origin yet it stood, albeit crookedly, for some 70 years - and for 30 of these with no maintenance while some other more substantial buildings have decayed away. Inside are concrete platforms that would originally have supported tanks containing warm and cool water, brine for cooling the nitrator while the reaction was taking place and also of course, tanks for mixed acid and glycerine. The acid tank was in a separate compartment at one end. Areas used for acid storage are easily identified by the hard bricks used and by the fact that a bitumastic substance was used as jointing instead of mortar. Between the Blackhill reservoir and AB2 stands a large brick building which contained large mixed acid tanks installed during the Second World War to supply the whole Nitroglycerine Hill.

Chapter 8

THE SECOND WORLD WAR

By 1939 the RNCF was more or less at its peak with its new Schmid continuous production nitroglycerine plant in full use and with the Picrite Factory – built for the production of flashless cordite – almost completed. During the Second World War no fewer than 1,100 air raid warnings were sounded but, in spite of this, production was stepped up to about 16 percent above the normal maximum for the factory. This achievement speaks volumes for the efforts made during those critical times by the employees at Holton Heath.

During the Second World War the guncotton nitration facility was extended as a part of the effort made to increase cordite output and as soon as production started in the new Nitration House it was noticed, during routine analytical checks, that the nitrogen content had fallen below the required level of 12.2 percent which indicated incomplete nitration of the paper. This discovery led to some intensive detective work and ultimately the explanation was found in an unexpected quarter. To find the answer it proved necessary to investigate the training of a new chargehand. It transpired that he had learned the process from the chargehand in the main guncotton nitration house who knew well enough that the paper scrolls had to be dipped and raised – in and out of the mixed acid – twelve times before the nitration was allowed to proceed undisturbed.

This repetitive dipping was carried out in order to ensure that the acid covered the entire area of the scrolled paper and no air pockets were present which would reduce the efficiency of the nitration process. In order to be sure that the job was done properly, the old chargehand used the dodge so often employed by cricket umpires to check off the balls in an over, although instead of counting pebbles or coins – neither of which would have been allowed in the guncotton factory – he counted the windows, of which there were twelve running along one side of the building. The trainee chargehand was keen to succeed and he learned the job thoroughly from top to bottom and, no doubt, all would have been well if the new extension had been fitted with twelve windows. Unfortunately there were only six.

In addition to their normal heavy workload nearly all the employees, like most people at the time, were also involved in other activities designed to assist in the prosecution of the war, and for many men this meant a commitment to the Home Guard. The No.5 RNCF Company Home Guard, 7th Dorsets, was led by Major Sparks who was Officer-in-Charge of works. The second-in-command was Captain W Coombes, who in everyday life was foreman in the Works Department and was deeply involved in the large scale construction that was undertaken in the lead up to the war. Before joining the RNCF in 1928, he had been a regular soldier and doubtless this experience equipped him well for his Home Guard duties. The Company was 185 strong with, in addition, four lieutenants, one CSM, a CQMS, six sergeants, eleven corporals and 140 privates. These figures applied specifically during 1942 but the company was presumably at about this strength throughout the war.

The Home Guard performed a variety of tasks, not the least of which was the 24-hours a day manning of the two decoy sites 'HH1' and 'HH2' situated at Arne and Wareham respectively. These sites were designed to lure enemy bombers away from the factory and the first was built at Arne in 1941 in the remarkably short time of six weeks.

This period included not only the construction but also all the essential preliminaries such as surveying, preparation of drawings and the clearing of the site. All the buildings were of tubular scaffolding and some were covered by metal sheets salvaged from scrap metal drums. All the facilities at the site were operated remotely from an underground shelter. The site was manned by two members of the Home Guard, one of whom patrolled the fence above ground while the second was in the control trench. On receipt of a signal via the direct land-line from the factory, simulated incendiaries could be detonated. Fires were made by burning fuel oil and adding water in such a way that an impression of a blazing inferno with dense smoke was presented to observers from above. When it seemed appropriate, oil drums full of waste cordite could be exploded and in this way the deception was completed to such effect that, on the morning following one intensive night raid, no fewer than 120 craters were found on the land. The following year, a similar decoy site was

Above: An interior view of the control trench – an underground bunker – which provided emergency communications and bomb plotting equipment. The ladder at the back leads up to a protected observation turret. To the left are the public address equipment and a chemical toilet. The tunnel leading to this room was provided with seats and could be used as a general shelter.

Right: The emergency exit and observation point which served the control trench. This position would have enjoyed a wide view over the factory at the time of the Second World War, but due to the subsequent growth of trees and bushes, the view is now much restricted.

constructed on Decoy Heath at Wareham to the west of the factory in an equally short period of time. This construction included the erection of 5,000 feet of boundary fencing.

The manning of these sites must have been a very unpleasant job, ranging from the tedium of quiet nights to the excitement, horror even, of those nights when to be successful in one's allotted task was to expose oneself to a blitz. Such methods were also employed to divert enemy attention from British cities. Known as 'Q' sites, these often employed tricks of lighting to simulate the view which might be expected from the air.

Air raid warnings for the Holton Heath area were immaculately recorded in a diary kept by a Mrs Gover, who lived at Ridge – a couple of miles north of the RNCF decoy site at Arne – on no fewer than 213 days between 04:00h on Friday, 14th June 1940 and barely more than a year later on 17th June 1941. That first entry recorded 'an air raid warning – nothing doing'.

It turned out that there were warnings on most days and it was quite common for there to be several. On Saturday, 5th October 1940 there were, for example – but not unusually – three warnings. It is presumed that most of these events were either false alarms, in that they turned out to record the appearance of enemy aircraft in the region of Holton Heath but en-route for other target areas, or failed to do serious damage. However, all of them would have caused disruption to a degree and it was surprising to find that the warnings were so very frequent, occurring as they did on nearly six days out of every ten during this period.

Short additional notes occur here and there throughout the diary. On the morning of Thursday, 20th June 1940, during an attack between 00:25-03:00h, Mrs Gover recorded 'damage done'. On Friday, 2nd August 1940, she 'went in shelter for first time'. On the evening of Thursday, 8th August she wrote 'Spitfire plane come down opposite Wha--- (?)'. A 'Plane brought down at Arne' appears adjacent to the entry for a raid on 27th September. Monday, 7th October was the occasion of another raid between 03:40 and 04:40 and this apparently resulted in another 'Plane down on Heath'.

On 28th November 1940, a 'British Plane in Slepe Bog' was the major feature of a raid occurring between 01:50-05:35h.

During 1941, on Tuesday, 18th February, she records somewhat mystically, 'Bell tolled 17:25h'. A fairly important raid seems to have occurred on Friday, 11th April when enemy aircraft 'Dropped nine bombs on Heath'. On Sunday, 20th April though there was one warning only, but it was Hitler's birthday of course!

However, there were air raids that occurred beyond the 1940-1 period recorded so carefully by Mrs Gover. In a memorandum from the Director of Local Defence in 1942, several raids were described during which many lives and much property was saved – together with essentially uninterrupted war work – as a consequence of the successful installation and operation of decoys at various Naval establishments. A raid directed at the RNCF was effectively deflected to the decoy site at Arne on the night of 3rd/4th June 1942 despite the expectation that the target would be found as a result of the clear visibility that prevailed. The successful decoy was largely due to the well-timed lighting of QF fires by the Home Guard on duty that night. It seems that maybe sixty enemy aircraft were involved but it is certain that in excess of 112 high explosive bombs were dropped together with somewhere around 5,000 incendiaries. Almost all fell around the decoy or around heath fires started by the enemy's own action and there were no casualties and no damage done.

Although we know that there were over 1,100 alerts during the Second World War, it has been difficult to establish just how many raids actually resulted in the dropping of bombs either in, or near the factory. However, it does appear that a single plane dropped sticks of bombs over the factory on three successive nights. On the first occasion, incendiary bombs fell in the area from the power station to the car park outside the main gate. Eight bombs fell by the boiler house and the scars are still visible on the walls. On the second night, bombs fell inside the factory to the north-west in the area of the so-called Cordite Village, in the garden of Heathwood Garage and across the common to Organford. On the third occasion, bombs fell in the higher Lytchett area, well away from the factory.

A bomb which landed in the car park outside the main gate left a very large hole, big enough to accommodate two buses it is said. It also damaged a lot of bicycles! Another similar bomb fell near the tennis courts to the north of the General Offices but in neither case were there any injuries, nor was there any significant damage.

Above: Lt Woodrow RNVR, with a substantial team, are justifiably pleased to recover an unexploded bomb from marshy ground. A substantial amount of work was required, with wooden shuttering necessary to prevent ingress of soil during excavation.

An unexploded bomb, which fell on marshy ground near Heathwood Garage towards the Bakers Arms and on the same side of the road, was finally recovered and defused by the factory bomb disposal officer Lt Woodrow, who was assisted by Major Sparks and a gang of men. The bomb was apparently difficult to recover because as they dug down to it, so it sank even deeper into the boggy ground.

In April 1944, a 5Kg phosphorous bomb fell within the factory but failed to detonate. This bomb was steamed out and was on display in the main office building for a number of years. Potentially the most dangerous attack resulted in an 11Kg oil incendiary bomb falling immediately adjacent to the southern end of the wash water settling house in the nitroglycerine complex. The bomb fell at 10:10pm on Thursday, 29th August 1940. The building, made of wood, caught fire but this was quickly extinguished and the bomb smothered as a result of a very prompt and courageous action by a number of people, including a chargehand from E3 called Cobb, a duty chemist called Silvester and the rest of a fire team led by Walt Dominey.

Right: A Second World War photograph of a fire-fighting team with trailer pump. The officer in charge is Capt K F D Ackland and the ladies (left to right) are
Gwen Harris (Lab)
Jean Maidman (Office)
Molly Holmes (Office)
Margaret Yeatman (Lab)
Joyce Green (Office).

Left: A fire-fighting team poses for the camera either before or after a soaking from their hoses whilst carrying out practice with the pumps. Some of the faces will be recognised from the picture above.

Above: A Second World War aerial photograph of the site kindly provided by the Luftwaffe, and taken in 1940. Although not easily seen in this copy the areas of the factory were correctly marked, thanks to some degree one suspects, from intelligence gathered during the installation of the Schmid unit just prior to the outbreak of war.

Right: An anti-aircraft installation, probably located on the heath surrounding the factory.

Left: A view towards the nitrocellulose factory, showing the installation of a typical air-raid shelter or refuge. Many on site were of similar design, being simply concrete tubes, later to be covered with earth. This design was not exclusive, as some ground-level walk-in shelters with reinforced roofs were also provided.

Right: This building, wash water settling house 'D1', had a narrow escape when a German incendiary bomb was dropped nearby. The building was saved by the staff fire-fighters. Their action was not without grave risk, since the building contained nitroglycerine.

The heaviest raid of the war, as far as the RNCF was concerned, was that referred to earlier, on the night of 3rd June 1942. On this occasion, the pathfinders were much too close for comfort and the Superintendent, Captain Morgan, sent up the decoy at Arne. Incidentally, Captain F T de M Morgan was a member of the Tredegar family of Light Brigade fame and was born in Gibraltar. He was a good and much respected Superintendent. The decoy sites were activated by giving orders via a direct telephone link from the factory.

The message 'Press button A' led to ignition of incendiary devices whilst 'Press button B' set fire to a structure involving the use of a mixture of diesel fuel and water, which gave the impression of a substantial conflagration. When required, up to 900lbs of waste cordite, stored in drums, could be exploded at once. The decoy site must certainly have justified the time and effort expended in its construction on this night alone, for when Lt Woodrow visited Arne on the next day, he accounted for 206 bombs and many more must have fallen into the waters of Poole Harbour.

The camouflage of bomb craters was also of the greatest importance in order that bomb damage should be hidden from aerial reconnaissance. The Works Department at Holton Heath devised an ingenious technique for this purpose which involved the use of camouflage netting supported by ⅛ inch wires strained over a preformed structure of scaffold tubing and ¾ inch wire cable. These devices were constructed beforehand and rolled up for easy transport to the damaged areas.

As would be expected, the defence of such an important part of the British war machine was also provided with anti-aircraft gun support in a ring around the Holton Heath site, and it is with thanks to Fred Ryder and C W 'Alex' Alexander that we have the following details. At the outbreak of the war, Holton Heath was defended by an outer circle of four AA sites, the so-called 'heavies', at Upton, Slepe, Northport (Wareham) and Arne, each initially equipped with two 3 inch First World War vintage guns (firing 16lb shells) without either predictors or height and range finders. Later of course, more modern and sophisticated AA equipment became available including 3.7 inch and some 4.5 inch guns.

A circle of smaller calibre guns, known as the 'lights', formed an inner ring and was begun with a few Lewis guns with the occasional 20mm Oerlikon and Hispano guns thinly distributed around the site. Later, 40mm Bofors guns were installed and they, like the others, were sited in shallow pits within a breeze block wall surrounded by an earth revetting piled against the outside. By 1944, operational sites were established at Sandford House Towers, East Holton, Holton Tower, Black Hill, Worth, Arne, Round Island, Fitzworth Point and Shipstall Point.

The operation of both heavies and lights was controlled from the Gunnery Operations Room at Lytchett Manor. Evidence of the presence of the heavy sites is limited to Arne where the remains of the command post and four gun emplacements are still to be seen.

These anti-aircraft defences were manned, not by the RNCF Home Guard but by a unit of the Territorial Army, formed at the RNCF early in 1939. This unit was known as the 77th Light AckAck Battery of the Royal Artillery and comprised some 150 staff volunteers commanded by officers with some military experience. These included Major Sparks, Chief Mechanical Engineer Dalton and Armament Supply Officer Keightley. These 150 men were also individually equipped with Lee Enfield rifles, some dating from the First World War. The men would continue to work at their regular jobs until a red alert sounded when they would down tools and rush to their gun positions, except at night when defence depended upon darkness and the Home Guard with their decoy fires. In fact, this group was never called upon to deal with any raids right up until the end of May 1940 when this element of the defence of the factory was undertaken by regular troops.

These various devices, coupled with anti-aircraft activities described above, were so successful that throughout the entire period of the war very few bombs fell within the factory fence. Those that did fall did very little damage. Each section had its own fire team and during wartime even greater attention must have been paid to fire fighting. The women workers in the factory were well able to handle trailer pumps and the other paraphernalia associated with fire fighting and it may be guessed that as long as it was only practice then these activities made a welcome break in the daily routine.

Air raid shelters – shown as 'refuges' on the site plans – were constructed at the beginning of the war from reinforced concrete components with an elliptical cross-section which were partially below ground level. The soil excavated for this purpose was thrown over the top. Some of these refuges still exist and in the mid-1990s were remarkably dry and in good condition inside, some still with their wooden benches running along either side of a duck-boarded central aisle. An amusing incident arose when a developer contacted one of the authors as he was rather worried about possible contamination of an underground 'tank' which he had discovered on his newly-acquired plot. This turned out on inspection, and to the relief of the developer, to be yet another air raid shelter!

Above: A twin Lewis gun emplacement, typical of the anti-aircraft defences used at Holton Heath in the early years of the Second World War.

The isolated rural site of the RNCF and its regularly shaped earthworks – characteristic of an explosives factory – must have presented a very considerable camouflage problem, as must the railways which from the air would appear as 'pointers' to the buildings that they served. The extensive tree planting of 1927 certainly helped the installation to blend into the surrounding country. The original intention was that such plantations, strategically placed, would help to reduce the effects of blast in the event of an explosion.

The exercise was on quite a large scale with the records showing

Above: A group of women fire-fighters on exercise with a trailer pump alongside the figure head.

that in all, some 27,500 trees were planted. At least twenty different species were used, including chestnut – both Spanish and horse – oak, ash, poplar, beech, elm, holly, birch, sycamore, spruce and lime. Many of these still thrive, some making magnificent avenues along former paths, even though the paths themselves have disappeared under a carpet of leaves and grass.

Apart from the natural cover, supplementary camouflage was absolutely essential and all the usual techniques were employed, including the use of netting to break up regular and easily recognisable shapes and of black, green and brown camouflage paints on roofs, walls and concreted paths. Textures were added to smooth surfaces with granite chippings and additional trees, shrubs and heather patches were planted to help break up the green turfed area which covered the earth traverses between the buildings. When new buildings were erected, coloured facing bricks were used to good effect and these, together with painted walls, could be seen for some time after the war had ended.

During the course of the war, the RNCF had the doubtful privilege of receiving the attention of Lord Haw Haw and on more than one occasion he claimed that the factory had been completely destroyed.

Of course, not all wartime memories of life at RNCF are sombre ones and the illegal use of waste solvents as alternative fuels for cigarette lighters, has been recalled with glee by a non-smoker employed at RNCF during the Second World War. The particular reminiscence which was found so amusing had to do with an unsuspecting worker who charged his lighter with benzene just before coming off shift and going to the cinema. At the end of the show, during which he presumably enjoyed a smoke, he emerged into the cinema foyer with a face as black as the ace of spades as a result of the extremely sooty flame which is a feature of burning benzene.

On another occasion during 1942, it was discovered at the North Gate that someone had lost a gold watch in a changing room. It was decided that at the end of that shift at 10:30pm, everyone would be searched as they left the factory. It naturally took a very long time to search hundreds of people, but the task was ultimately completed although unfortunately the watch was not found. Looking back we can only imagine the expressions on the faces of the Royal Marine Police as they walked into the clocking station and found the floor littered with thousands of sheets of toilet paper, each one separately bearing the words 'Government Property'. Many will know that toilet paper was a scarce commodity in those wartime days and for this reason it was commonplace for it to be pilfered. Incidentally, the special workers' train to Christchurch, due at 10:45pm, was held in Holton Heath station until the search was finished, as also was the train to Swanage.

One dodge that flourished for some time was for a selected workman to clock on as usual and to collect betting slips from around the site. He would then be let out of the site via a loose railing (a 'King Dick' adjustable spanner being reserved for the purpose) whereupon he would catch the train from Holton Heath to Poole, where the morning was spent in placing the bets. He would then return by the same route.

The people at RNCF were of all kinds, but, whether they were using waste solvent in lighters, glycerine to soften work-hardened hands, or cordite trays as garden fences, they seem to have been a remarkably resourceful lot and never was this quality more impressively displayed than on one occasion when an assistant foreman in the picrite section was found in his office asleep with his head in his hands. On being somewhat rudely awakened and accused of being asleep while on duty, he promptly denied the charge and insisted that he had, in fact, been praying for the safety of his men.

Although the factory itself escaped serious damage from aerial attack, there was one occasion when it very nearly lost a train load of its workers. The train was the 5:19pm from Holton Heath to Christchurch and the incident occurred on the 21st March 1941. The train, crowded with young women workers – and known locally as the 'Glamour Puffer' – was crossing Rocklea Bridge on the way to Poole when a plane dived on it and dropped a stick of six bombs. Fortunately, none of these struck directly either the train or the permanent way, although there were very near misses indeed and windows throughout the train were broken and much superficial damage was done to the rolling stock. Miraculously no lives were lost and neither were there any serious injuries although, naturally, many were badly shaken.

Special trains were arranged to convey workers to and from Holton Heath, starting from Christchurch and Swanage. All shifts – first, second, third and day – were provided for. It was at about the time of this bombing incident that a particular rake of rolling stock was used for certain of the special factory trains. It consisted entirely of very old first-class, non-corridor coaches originating from the South Eastern & Chatham Railway and was of the kind known to railway buffs as 'birdcage' stock.

This rake, starting from Christchurch, was used to work the first shift train leaving Poole at 5:31am. It then ran to Wareham to turn around and returned empty to Christchurch to work the day shift train calling at Poole at 7:52am. The stock was then left at Wareham until required to work the 5:19pm from Holton Heath to Christchurch. It then worked the night shift train in, leaving Poole at 9:12pm and going on to Wareham from whence it returned as the 10:45pm from Holton Heath to Christchurch where it remained overnight.

John Gray recalls an incident which occurred on one Summer evening when he was travelling on the 5:19pm. The train had just started when there was a shout and the train stopped while the van was still at the platform. It left a few minutes later. The cause of the stoppage was an American soldier, perhaps a hobo in private life, who was riding on the buffers and was noticed when the train started.

Bombs were not the only thing to be dropped from the skies in those wartime days, for it was during the early part of the Second World War, in 1940, that a parachute with a horses saddle suspended from it was seen to fall from a German aircraft and descend into the factory. This event was a complete mystery at the time and as far as is known, remains a mystery to this day. However it did provide a talking point and for a time this improbable combination was on display in the main office building.

Later in 1942, the RNCF had a very lucky escape when a Flying Fortress, which had run out of fuel on the way back from a raid, made a pancake landing in fields on the factory side of the road towards Lytchett Minster. The aircraft was carefully guarded during the time it took the Americans to bulldoze away hedges and make all other arrangements necessary to fly it away.

The Home Guard Company held their stand-down Annual Dinner at the British Restaurant in Poole on Saturday, 2nd March 1946. A toast to the RNCF Company was proposed by Captain Kenneth Francis Dyke Ackland RN, who was Superintendent at the time and Major Sparks responded. Entertainment was provided by Messrs Elmes, Benjafield and Boyce.

Chapter 9

SOCIAL ACTIVITIES

Above: Although not taken at the factory, this group photograph of RNCF workers shows clearly the protective clothing in use at that time and the dipping forks used in the production of guncotton.

There was a very active social life based on the factory. One of the earliest major events to be recorded in the local press was a sports meeting held in Poole Park in June 1924.

A number of open events were held and it appears that many of them were won by representatives of the Royal Tank Corps at Bovington. The factory competition included a tug-of-war which was won by the cordite section, a sack race and a one-mile gentlemen's cycle handicap race as well as a wide range of more conventional athletic events. The prizes were distributed by the Superintendent's wife, Mrs Desborough. The Parkstone division of the St. John Ambulance Brigade was in attendance, we are told, and their new motor ambulance was brought into public service by conveying one lady with a back injury to hospital and another – taken seriously ill – to her home in Bournemouth.

Although the sports meetings may not have become a regular event, the first of what was to become an annual series of flower shows was held on Saturday, 10th August 1929. These shows were held at the factory and were a much enjoyed event for at least nine years. The *Poole & East Dorset Herald* said that 'A highly successful show of flowers, fruit, vegetables and honey was held on Saturday at the RNCF, Holton Heath, by the staff. The show owed its inception to the kindly thought of the Superintendent' .We are also told that 'The show was restricted to members of staff and their families and some 500 persons attended. General approval of the arrangements was expressed by all present.' In all there were 190 entries by some 40 exhibitors and these were given 'a most searching examination' by the three judges. The prizes were presented by the wife of the Superintendent, Mrs Hammond, and the produce was sent to the Cornelia Hospital, Poole.

There were fringe attractions too. During the afternoon a display was given by two of the factory's fire teams, which included six man drill, three man drill and one man drill. A very strong tug-of-war team from the Works section was present to challenge all comers and easily proved to be the master of a scratch team from the Cordite section.

The Holton Heath Tennis Club arranged a tournament for the occasion and this appears to have been very successful with thirty-two players competing. Music was provided throughout by The Radio House, Poole, by means of a 'travelling broadcasting van'.

The Flower show of 1937 was the ninth and was very successful with nearly 1,000 people attending. A record number of entries were received and the quality of the exhibits was said to be much higher than that of the previous year. The prizes were presented in the traditional way by the wife of the Superintendent, Mrs F T de M Morgan.

Other attractions included 'rhythmic dancing by the Rhythmic School of Dancing' and exhibition boxing by members of the Boys' Club as well as an exhibition of working bees, children's racing, various sideshows, clowns and performing dogs! Another tennis tournament was held and the Superintendent's garden was open to visitors. Tug-of-war competitions were held, including one for ladies in which first-shift guncotton beat second-shift guncotton in the final. In the men's competition the Works section won again in both the heavy and the light divisions.

A special demonstration was put on in the afternoon by the Air Raids Precaution squads, who showed how quickly anti-gas measures could be taken in the event of an airborne mustard gas attack. The squads were said to have coped efficiently with large numbers of casualties and to have quickly decontaminated and cleared the area, 'enabling the flower show to be continued.' A number of entertainments were held at the factory during the war, many of them in a mess room which was demolished some years ago, after being briefly a part of Overhill Engineering Ltd, just off the main Poole-Wareham road. Several visits were made to Holton Heath by ENSA parties but it is understood that they were not always well received, as the standard of entertainment was not always very high.

Apart from this, it was largely a matter of self-help and entertainments of various kinds held at the factory were organised by the staff themselves. N A 'Henry' Ford, who was mainly concerned with passive defence (PD) in the factory at this time, produced a number of shows and plays. One of these, which appeared early in 1945, was called *Cordite Flashes* and was described as 'two hours non-stop variety'. Featured in this particular show were Dan Steele (monologues), Charlie Foster (singing comedian), Arthur Tilsed and George Haberfield (comic duets), and Vivienne Maidman with Male Voice Chorus in a *Song Scena*. William Haskell sang ballads and Professor Elmes and Doc Benjafield were involved, in some way, with *The girl who lost her head*. Peggy Walton, Ron Abbott and John Withey appeared in *The Haunted House* and Alan Wilkins ARCO, conducted the RNCF Theatre Orchestra.

MAKING CORDITE
COTTON
GUN
MIXING ACID
MAKING GUN COTTON
LAB.
CLERICAL STAFF
COST
MIXING
ENGINEERS
SCIENTIFIC STAFF
RNCF
WORKS
ROLLING
STORES
R.N.C.F.
PRESSING
37.

Left: Drawn in 1937 as a Christmas card for the RNCF by Leonard Charles Ernest Howley, a well-known personality at both RNCF and RNPF Caerwent. At the time he was a nitroglycerine chemist.

Chapter 10

THE PRODUCTS & HOW THEY WERE MADE

A brief history of cordite and its precursors

The invention of smokeless powders – including cordite – as alternatives to gunpowder, which had been known for more than five centuries, was presaged by the discovery in 1846 of nitroglycerine. The chemist responsible was an Italian called Sobrero who obtained the material by the action of nitric acid on glycerine. In 1863, the celebrated chemist and engineer Alfred Bernard Nobel began to use nitroglycerine as a blasting explosive. It was called 'Nobel's blasting oil' and was rather dangerous to handle because of the capricious occurrence and great violence of its explosion. However, three years later, the discovery by Nobel that Kieselguhr – a diatomaceous earth – would absorb up to three or four times its weight of nitroglycerine turned out to be the key to the problem. Experiment showed that the Kieselguhr stabilised the liquid nitroglycerine, making it safer to handle and to transport, but it also conferred the additional benefit of making the mixture more convenient to detonate than the nitroglycerine alone. This material was patented as dynamite in 1867.

Meanwhile, the development of another extremely powerful organic explosive was nearing completion. This material, which was to be called guncotton, was made by the action of nitric acid on cotton and the discovery had its foundations in work done by Pelouze in 1838 and by Schonbein in 1845. But it was not until 1865 that the practical difficulties of applying guncotton as a shattering explosive were overcome by Sir Frederick Abel. Attempts were then made to reduce its furious rate of burning to a level where it could be used as a propellant and ultimately, in 1886, a Frenchman named Vieille achieved some success with a solvent-modified derivative of guncotton, named Poudre 'B'.

However, it was left to the genius of Nobel to unite the two lines or research, starting from glycerine and cotton, and to show that the explosive properties of nitrated cotton could be tamed for propellant purposes by gelatinising the fibrous material with nitroglycerine. The discovery that the two most powerful explosives then known could be blended to give a slow burning propellant 'was so startling that it was received with incredulity, which soon gave place to extreme astonishment'. This resulting smokeless powder was named Ballistite and was patented in 1888.

At about the time of the discovery of Poudre 'B', the British Government appointed a committee to investigate and report on the subject of a suitable smokeless powder for the British service. Samples of ballistite and a number of other powders were obtained, but the committee was not satisfied with any of them and finally devised a new one, to which the name 'cordite' was given. The name derived from the fact that it was made in the form of 'cords'.

This work was conducted in Sir Frederick Abel's laboratories at Woolwich Arsenal and the material was patented in 1889 on behalf of the government in the names of Sir Frederick Abel and Professor (later Sir) James Dewar, who were members of the committee. Nobel claimed that his ballistite patent of 1888 was infringed through the production of cordite, which was commenced at the Royal Gunpowder Factory, Waltham Abbey, later in 1889, or possibly early in 1890. Certainly cordite was, like ballistite, essentially an admixture of nitroglycerine and guncotton but there were a number of differences and these proved to be important enough to ensure that the ensuing lawsuits ended in favour of the British Government.

One of the most important differences between ballistite and cordite was that the latter used acetone as a solvent to assist in the incorporation of the nitroglycerine with the guncotton. A second important difference was that cordite had a proportion, amounting to 5% by weight, of mineral jelly, similar to Vaseline, included in the formulation. The mineral jelly was originally included in the hope that it would lubricate the barrels. This does not happen because it is entirely consumed in the explosion, but it does have another very desirable effect. The presence of the jelly reduces the rate of erosion of the barrel by lowering the temperature of the explosion. This early product was known as Cordite MkI but, despite the presence of mineral jelly, it still caused severe wear in gun barrels in the Boer War. As a consequence, the composition was modified to reduce the temperature still further by decreasing the proportions of nitroglycerine from 58 to 30% and by increasing the guncotton concentration from 37 to 65%. The new version was known as Cordite MD and it was this material that was produced at Holton Heath for Naval use during the First World War.

Intermediates

The major end-product of the RNCF was the propellant cordite of course but, before this material could be made, there was a requirement for a wide range of raw materials and a number of these were also made there in 'factories within a factory'.

In particular, the bulk manufacture of explosives requires the use of vast quantities of pure, concentrated sulphuric and nitric acids and a brief description of how these materials were manufactured will be given. A description of the process involved at RNCF in the production of guncotton (or nitrocellulose) and of nitroglycerine will also appear. Also described is the production facility for picrite,the component essential to the production of flashless cordite which was so extensively used in the Second World War. The final manufacturing process described is that for the high explosive tetryl, which was also made at Holton Heath from the beginning

The production of acetone was also an important feature of RNCF activities until the late 1920s when a change from one form of cordite to another meant that it was no longer required in the same large quantities. However, due to lack of detailed information on the exact processes involved, as much as is known has already been presented. It will be noted that glycerine was one component not manufactured on site. This is because glycerine is commonly available as a by-product of soap manufacture resulting from the saponification of vegetable oils, and so presumably the Admiralty found a ready and adequate supply from the commercial sources already in existence.

Sulphuric Acid

The raw material for the production of sulphuric acid was iron pyrites, familiar as 'fool's gold', which was imported by sea to Poole from such places as San Telmo in Spain and from Bjorkassen, Norway. To casual observers the heavily laden vessels alongside Poole Quay appeared to have virtually empty holds as a consequence of the very high density of the pyrites. The cargo was transferred from the quay to the factory in railway trucks after the material had been sampled.

The specification for the pyrites stated that the weight to be invoiced and paid for would be that of the pyrites less the moisture content which was to be ascertained at Admiralty expense on delivery in the presence of the contractor's representatives. The price per ton was assessed on the basis of sulphur content. This was to be estimated by both the Admiralty and the contractor and the results obtained were to be exchanged on a pre-arranged date. If the two figures differed by less than 0.5% then the mean of the figures would form the basis of the invoice but if the difference was greater than this a check sample was then to be sent to an independent firm of analysts (at one time Messrs Riley, Harboard & Son of London or Rudd & Thompson of Port Talbot).

The umpires assay was taken as final and binding on both parties if it fell between the results obtained by Admiralty and contractor. If the umpires figure was outside, then the figure nearest to that of the umpire was taken as the settled assay and the cost of the independent analysis was borne by the party whose result was the furthest from that of the umpire. The residual ash from the pyrites was bought back by the contractors and sold to iron works.

The sulphuric acid factory had a pyrites store which held 6,000 tons situated close to the main railway line but, in addition, there was a reserve store at Rocklea where a further 4,000 tons were held. This meant that the factory had a total pyrites storage capacity of 10,000 tons, which was equivalent to at least eighteen months supply.

Sulphuric acid was produced by the so-called 'contact' process in a Tentelev plant. The pyrites was burned in the plant at a maximum rate of 20 tons a day to give sulphur dioxide gas which, as many will know, is extremely irritating to the eyes, nose and throat. This gas was then drawn, together with air, over a platinum catalyst which was used in a finely divided form on an asbestos support. The 24lb weight of platinum in the Tentelev plant facilitated the interaction of the two gases to give a third gas, sulphur trioxide, which was dissolved in concentrated sulphuric acid to give oleum which was then diluted with water to strengths appropriate to the various process in which it was later to he employed. By 1939 this acid was being produced at the rate of 150 tons per week.

During the Second World War, when Norwegian pyrites was not available and Spain was a doubtful country, a Glen Falls sulphur burner was installed in the sulphuric acid plant and Louisiana sulphur was supplied by the USA under the lease-lend arrangement. Later still, on the cessation of hostilities with Italy, sulphur came to RNCF from this source. Some of it must have been carried on an old fruit boat because on several occasions lemons were found in the feed to the sulphur burner.

The working conditions were generally unpleasant in the acids factory. However conscientiously the plant was maintained leaks were inevitable and the atmosphere carried large concentrations of sulphur dioxide. A member of the acids laboratory staff recalls that, as a young assistant in 1937, he was required to collect acid samples for quality control purposes. The atmosphere within the factory was so unpleasant that he was in the habit of taking a deep breath just before entering and not breathing again until he emerged with the sample.

He recalls how the huge double doors into the sulphuric acid factory were always kept open, even on the coldest days of winter, presumably because even the cold was preferable to the effects of the choking sulphurous gases. In fact, keeping these doors open probably reduced the very high noise level which was also a feature of life in the acids factory. The noise was generated by the huge fans which drew the mixed gases over the catalyst. It is understood that, as a result of the presence of sulphur dioxide in the saliva, an abnormally high incidence of dental caries was to be found among this group of workers. By way of compensation, it is also said that on the whole these workers had beautiful complexions which may have been attributable to traces of arsenic compounds which are always associated with iron pyrites. This is an interesting story which ought to be true even if it is not. One can imagine a potential recruit being asked what he would like to hear first – 'the good news or the bad'!

It was interesting to note that – certainly until the mid-1990s – the area once occupied by the acids factory was the only land which had not fully recovered from its former industrial use, although no works had been carried out on the area since the factory was dismantled. Trees and shrubs have re-grown to some extent, but grass cover is still sparse. This is probably not due to the action of the acids, but more likely the fact that the fuel to heat each factory process was gas locally produced. In addition, a large amount of coal was stored in the area for producer gas retort-firing and it is suspected that it is coal by-products and waste that have inhibited the regrowth, much as they did at the site of the old Poole gas works.

Below: A view of the Tentelev plant, taken from the north.

Left: The Tentelev louzal pyrites store. This was held in a large building close to the main railway line and at the Eastern end of the factory. The material was of a high density, resulting in a high ratio of weight to volume.

Right: The pyrites, being a mined material, was initially screened in this rotating sieve to remove the bulk of unwanted material such as gravel.

Above & opposite: Two views of the Tentelev Lurgi furnaces in which the pyrites was burned in order to release sulphur dioxide in the first stage of sulphuric acid manufacture.

Above: Gas coolers in the Tentelev plant. The hot sulphur dioxide from the furnaces was cooled prior to purification in the 'scrubbing' towers at centre. The purified gas was then applied to the platinum catalyse. The reaction with the catalyst produced sulphur trioxide.

Above centre: Quartz filter feed apparatus in the Tentelev plant.

Below centre: Scrubbing and drying towers.

Above right: Absorption vessel.

Below right: Oil filters and turbo blowers.

Above: Tentelev gas cooler interchanger and contact vessel. The large vessel housed the platinum catalyst. The Sulphur trioxide emerged from this plant and was then dissolved in concentrated sulphuric acid to give oleum - later to be added to water to produce sulphuric acid to the required strength for the nitroglycerine / nitrocellulose nitration phase. A similar process could be used to 'work up' weak recovered acid from previous use, following de-nitration.

Above: Tentelev arsenic testing bench.

Right: Temperature recorders.

Nitric acid

The nitric acid was made by the action of heat on a mixture of sulphuric acid and sodium nitrate, or Chile saltpetre, contained in cast iron retorts, or stills, of which there were twelve installed at Holton Heath. These twelve retorts used about four tons of gas coal each day when the plant was in full operation. The output of nitric acid was about 100 tons per week in 1940. A by-product of this process was a material called nitre cake and at peak output as much as 150 tons per week was produced. This material was used in the manufacture of *Harpic*, the well known domestic cleanser. About 50 tons a week was sent to Wilmington, near Hull, for the use of *Harpic Ltd* and the remainder was dumped. There were some old railway trucks kept especially for this traffic on account of the corrosive nature of the material. Some of these were of London, Brighton & South Coast Railway origin.

Right: A view of nitrate skips at the nitric acid factory.

Below: The nitrate store at the acid factory.

Right: A cummer dryer in the nitric acid factory. Patented by F D Gummer in the USA in 1889, the apparatus was able to reduced the water content of granular material to less than 2% without excessively high temperatures.

Left: Nitric acid factory dry nitrate conveyor.

Above: A general view of the upper floor of the nitric acid factory, showing, on the right, a series of silica condensing coils.

Opposite: Fume absorption and acid collection plant in the nitric acid factory.

These pictures show the production of nitric acid by the action of heat on a mixture of sulphuric acid and sodium nitrate. The acid from the process or from recovery would be collected from towers which delivered the acid by gravity into the vessels underneath.

Opposite Absorption tower and stoneware receivers.

Right: One of the producer gas plants which supplied the required heat.
Centre: Nitric acid factory fume absorption towers.

Acids recovery

During the Second World War nitric acid was brought into RNCF from the Royal Naval Propellant Factory and ICI Ltd, where it was manufactured from ammonia. The old nitre stills continued to be used to make up any deficiency and the process was an economically attractive way of using old and dirty sulphuric acid. In addition to the manufacture of sulphuric and nitric acids, the factory was also responsible for recovering and revivifying spent mixed acids after they had been used. This was a most important aspect of the acids factory work and it was therefore equipped with a whole range of plant for these purposes, including de-nitration towers (to remove nitric from sulphuric acid), detetrylating plant, Kessler concentrators for sulphuric acid and much else outside the scope of this account.

It is worth noting however, that Kessler concentrators must have been formidable pieces of equipment, since the accident book records a high number of acid burns received when maintenance was in progress, including several occasions when workers fell into the interior.

Below: The acids used in nitration of cellulose or glycerine require a mixture of sulphuric and nitric acids. This picture shows the pumping and mixing apparatus which provided the correct ratio of acids.

Right and below:
These mixed acid tanks stored the combination of required acids prior to being delivered to the nitrating houses. The size of the holding tanks gives some indication of the production volume of which the RNCF was capable.

A very important function of the acids factory was to facilitate the recovery, concentration and re-use of waste acid from the nitration processes. This involved de-nitration to recover sulphuric acid and was achieved by a further distillation process. Traps were also required to check for and remove any remaining traces of nitroglycerine. Weak acid was recovered from the purification plant which was then worked up to the required strength.

Above: An interior view of the nitrocellulose section waste acid pumping house.

Left: Waste acid inspection apparatus.

Right: Denitration plant – waste acid feed boilers.

Left: A view of the acid outlets at the base of the denitration towers.

Above: Denitration towers.

Above: Concentrator feed tanks.

Above: Kessler concentration towers, used to 'work up' weak sulphuric acid to a required strength. These units required heat supplied blocally by gas producers.

Above: Sulphuric acid concentration cooling and clarification tank.

Left: Sulphuric acid concentration towers.

Above right: Sulphuric acid concentration plant coke filters.

Right: Local gas producers in the sulphuric acid concentration plant.

Cottrell plant

Recently acquired information revealed an interesting insight into the process of sulphuric acid recovery and concentration. This showed that a system known as the 'Cottrell plant' was employed to remove acid mist and, more importantly, toxic lead oxide dust, from exhaust via the fume vents. The system was patented by Mr Fred G Cottrell in 1907, who showed that it was possible to remove such material from an air carrier by the use of electrostatic attraction (a process later employed in fossil-fuelled power stations and still in use today).

Photographs of the plant show the equipment designed to produce the very high voltage required for the electrostatic capture and it is interesting to note that the rectifier used to change the alternating to direct voltage was of the mechanical synchronous type, effectively a rotating switch driven by a motor synchronised with the mains. Today, this transformation would be simply achieved by a few silicon diodes but, at the time of installation, no such convenient device existed. This precautionary air purification, along with the regular testing of sea-going effluent (even during wartime), well illustrates the care for the environment shown by the Admiralty. No such consideration could be given to an area a few miles along the coast at Studland Bay, where one anti-invasion technique was to set fire to the sea with pumped oil – the repeated test of which inhibited sea life in the region for some years after the conclusion of the Second World War. This extreme measure was thankfully never required, although it clearly needed to be effectively tested.

The Acids factory formed an administrative unit with the Guncotton factory and the directing staff of this unit consisted of a Chemist-in-Charge assisted by a Chemist. This unit was also responsible for the supervision and control of the Tetryl factory and the picrite factory.

Right: This picture shows the towers carrying the high voltage electricity supply to the Cottrell plant dust and fume collectors.

Far left: A view of the electrical control panel for the Cottrell plant. The synchronous rectifier is at right.

Left: This shows the low-to-high voltage oil-filled transformer. Thin leads are just visible leading from the two tall terminals via the horizontal insulators, down to the rectifier.

Above: A close-up view of the synchronous rectifier.

Guncotton

The first batch of guncotton was made at the RNCF on 16th August 1916 when cotton waste was used as the raw material. The procedure commenced on 7th August 1916 with the picking over by hand of cotton waste to remove solid foreign matter. After this the cotton was passed to a teasing machine which opened up the material and cut it so as to render it suitable for the acid dipping process. When the factory started production four such teasing machines were installed and available for use.

Immediately before treating with acid, or nitrating, the cotton was dried in a current of warm air over a period of about three-quarters of an hour to reduce the water content from a typical value of about 8% to about 5%. At this time there were at least three methods commonly used for the nitration of cotton but the one used at RNCF was the so-called Nathan-Thomson Displacement Process, in recognition of the Superintendent and Manager at the Royal Gunpowder Factory at Waltham Abbey who developed the process and where it had been in use since 1905.

The process was carried out in earthenware pans which were normally grouped in operating units of four. Each pan was about 3 feet 6 inches in diameter and 10 inches deep. The pans were each covered with an aluminium hood via which the fumes were sucked away. The four pans were connected by lead pipes to supplies of mixed concentrated nitric and sulphuric acids and also to a water supply. The bottom of each pan was covered by a perforated plate over a central opening half an inch in diameter and the whole pan was supported on an earthenware pedestal 20 inches high.

Mixed acid was run into the pans, the shape of which was such that 6001b of acid was above the perforated plate and 501b was beneath it. Twenty pounds of cotton were then introduced in small quantities at a time using aluminium dipping forks for the purpose. An upper perforated plate was then fitted to each pan in segments and a film of cooled water was run onto the surface of this plate. This water layer was intended to prevent the escape of further acid fumes.

In spite of these precautions, the atmosphere in a guncotton nitrating house was unavoidably laden with such fumes. Ramsey and Weston – in their *Manual of Explosives* published in 1916 – when giving advice on industrial poisoning amongst explosives workers, said 'workmen should never remain in the nitrating house, for instance, of a guncotton factory during the whole of the process as, in spite of the means adopted, and theoretically supposed to prevent acid fumes from rising from the pans, they are in practice usually much in evidence, and the nitration house atmosphere therefore becomes anything but conducive to good health.'

Guncotton factories were therefore very unpleasant places to work in at the best of times and very nearly unbearable at the worst. Bill Austin remembers working a particular night shift when they had taken over from a second shift which had, in some inexplicable way, used unfit acid. The result was that nearly all the 200 pans in the nitrating house started to belch red fumes as the nitrations became unstable and the temperatures of the reactions soared out of control. The earthenware pans broke and there was acid everywhere. Workers were desperately trying to run off as much acid as possible and then dashing outside for fresh air before they succumbed to the choking poisonous red nitrous fumes. Hosing down the entire house was the easy bit at the end of a long shift when all doubtless felt that they had earned the entire 27/- which it took them a whole week to earn in those early days.

In normal circumstances, the nitration process took two and a half hours and at the end of this time the acid was slowly run off at 171b per minute while water ran in on to the top perforated plate at the same rate. The acid was entirely displaced in three hours after which the water was allowed to drain through the guncotton and this took about one hour.

The next process involved boiling the guncotton with successive quantities of water in special vats made of wood and about 6 feet high by 8-10 feet in diameter. They were lined with an antimony-lead alloy which was reasonably resistant to corrosion by dilute acid and each was equipped with a steam heated coil and a central kier. Fifty-one such vats were installed at the RNCF guncotton factory. This was a crucial process in that it was designed to remove lingering traces of acid which, if left, gave rise to instability in the explosive.

After boiling, the guncotton was beaten in machines similar to those used in the manufacture of paper where a roller armed with steel knives rotates and reduces to pulp guncotton caught between these knives and similar static blades mounted on the bed plate of the machine. As the pulping process proceeded, the roller was gradually lowered closer to the bed plate.

The removal of grit, metal and other foreign matter was the object of the next process which involved the passage of a slurry of pulped guncotton in water along an aluminium grit trap 2 feet wide and 100 feet long. At the bottom, strips of blanket were held in position by lead clamps. Finally the suspension of guncotton passed over an electromagnet which removed any ferrous particles which remained. The bulk of the particulate contaminants sank to the bottom and was retained on the blanket traps. In later years these contaminants were removed by centrifuging.

The penultimate process involved washing in vessels called 'poachers' or 'potchers' which were equipped with powerful paddles which agitated the aqueous suspension of guncotton. The use of these machines for guncotton manufacture is a logical progression from their use in the paper making industry. After washing, the guncotton was allowed to settle and the wash waters were skimmed off. In this way insoluble impurities and contaminants with a lower specific gravity than water were removed, as were soluble materials released by the pulping process from the hollow centres of the nitrated fibres. This process was repeated at least three times and, in addition to the further purification of the product, offered an opportunity to blend together batches of guncotton, thus helping to ensure a uniform product with a nitrogen content of just about 13%.

The drying operation was then started with a centrifugal process after which the product left the guncotton factory and was sent to the drying stoves. These stoves were typical danger buildings. They were of light, wooden construction and were fitted with shelves upon which the product was placed. Fans blew air, heated to about 60°C as it passed over steam pipes, in through apertures at the top of the building and over the guncotton. Moisture-laden air was vented to the outside at the bottom. The stoves were normally lined with zinc sheet, soldered at the joints so that no cracks or crannies were left in which explosive dust could accumulate and so that at intervals, it could be thoroughly washed out. When guncotton was in the form of loose fibres it was placed in trays of copper wire, all of which were carefully earthed because of the danger of fires resulting from the build up of static electricity.

Drying of this fibrous guncotton was completed in about 24 hours, although if dried in a moulded form, as was sometimes the case, several days were required.

At the RNCF, cotton waste was used for just over a year until 30th September 1917 when a change was made to 'sliver', another cotton product. Sliver was used for almost eight years after which, at the beginning of April 1925, there was a move back to cotton waste. Linters was used to supplement cotton waste from 19th October 1925 until mid-December 1929. Linters is the short fibre material left on the American upland cotton seed after the long staple fibres or 'lint' have been removed. This linters material amounts to about 10% of the total cotton on the seeds but was of no great value to the textile industry in those days.

Left: The preliminary stage of guncotton manufacture: these machines break the bales of raw waste cotton prior to the processes of teasing, picking and drying.

Right: Cotton preparation: picking tables, teasers and driers.

Left: Cotton teasing and drying: this photograph shows a mechanical teasing machine in the foreground and a drying oven in the background.

Right: Fan driven conveyors delivering cleaned and dried cotton waste into delivery ducts for subsequent processing.

Left: Cotton waste preparation cooling rooms.

Nitrocellulose

On 27th March 1929 there was the beginning of a significant technical breakthrough when a non-cotton based cellulosic material was used for the first time in the form of shredded paper. From the end of 1929 onwards, paper was used exclusively in the production of nitrocellulose. Later on an important technical innovation led to the commencement of the production of nitrocellulose as a slurry, which was pumped for the first time on 21th November 1932 via a pipeline to the cordite manufacturing area. This important method of transport is still in use to this day.

It should be explained that the difference between guncotton and nitrocellulose lies in the nitrogen content of the product. Chemically they are both obtained by the nitration of cellulosic material and so they are both classed as nitrocellulose, but in military jargon the term nitrocellulose is reserved for material with a nitrogen content of less than 13%. Guncotton was used in the production of Cordite Mk 1 and Cordite MD, but nitrocellulose was used for Cordite SC.

This clever pumping scheme was devised in order to simplify the complicated processes involved in transporting nitrocellulose to the mixing houses. The essence of the scheme was that a suspension of nitrocellulose in water – of pre-determined concentration – would be pumped continuously through a ring main to the mixing houses, where a measured quantity of suspension could be drawn off into a calibrated tank at will.

The nitrocellulose pulp was stored in massive blending tanks at the pulping and moulding room in the guncotton factory and from there it was pumped by means of a centrifugal pump through a 4 inch galvanised iron pipeline to lead-lined wooden tanks 350 yards away at the intermediate house. The route taken is clearly illustrated on the site plan.

Communication between the pulping and moulding room and the intermediate house was maintained by a system of light signals. Each signal was visible in both houses regardless of where it originated and they were designed to regulate the pumping of nitrocellulose pulp to the intermediate house and the pumping of wash waters to the pulping and moulding room. The first of these operations was controlled by four signals: 'Send NC', 'Flush', 'Stop NC' and 'Pump Stopped'. The pumping of wash water was controlled by three signals: 'Washing Ready', 'Send Washings' and 'Washings Out'. A 'Not Ready' signal could also be operated from the pulping and moulding room and in addition a direct telephone line was also fitted between the two houses.

Similarly, electric light signals were installed between the intermediate house and the mixing houses, but in this case they consisted of series-connected red and green lights operated at the intermediate house. When nitrocellulose was being pumped through the ring main, the green light was shown and the pulp could be withdrawn as required but in other circumstances a red light was shown and when this happened, the cocks leading from the main were not to be opened.

The lead-lined tanks at the intermediate house had a total capacity of about 18,000 gallons of pulp and this was stirred continuously to keep the solid in suspension. The concentration of nitrocellulose in the pulp was checked at the intermediate house by means of a specially developed density test and adjusted by the addition of water so that 10lb of solid was contained in 11 gallons of pulp. The pulp was then pumped through a 4 inch galvanised steel pipeline which passed through all the paste mixing houses (the 'E' houses on the site plan) and formed the ring main with the return to the tank from which it was pumped. Pulp velocity within the ring was arranged so that it was turbulent and the suspension remained uniform throughout.

Having drawn off the required quantity of slurry (300lb) the water was filtered off and replaced with the correct amount of nitroglycerine-saturated return water at the right temperature, after which it was ready for mixing with nitroglycerine in the early stages of making cordite SC. The use of nitroglycerine-saturated return water, rather than fresh water, was dictated by reasons of economy which meant a saving of £800 a year in 1932, when only 33 tons of cordite SC were being produced each week. Nitroglycerine at this time cost about £70 per ton to produce. At this rate of cordite SC production, the nitrocellulose pumping scheme led to an immediate reduction in direct labour costs of £2,553 4s 0d per year. The scheme came into full operation on 28th December 1932.

Although the material was used for over eight years, shredded paper was not entirely satisfactory because it packed together in such a way that only a comparatively small area was available to the nitrating acid mixture. However, as a direct result of pioneering work by Dr R S G Knight at Holton Heath, the much more efficient scrolling process was used to prepare rolls of paper prior to nitration. The technique involved winding together a strip of plain paper, together with a second strip of the same width which had been crimped. The completed paper roll had an appearance similar to that presented by the edge of corrugated cardboard.

The great advantage of such rolls was that almost all the surface of the entire roll of paper could be very quickly exposed to the mixed acids, thus ensuring uniform nitration. This new process also did away with the need for manual dipping of the cellulosic material and the aluminium dipping forks became redundant. Special scrolling machines were installed and the move from shredded to scrolled paper was complete by 11th July 1938.

The paper, wrapped in hessian, normally arrived at Holton Heath in closed railway trucks. One bale contained six rolls of paper 7½in. wide, each weighing 40lb. Very large quantities of paper were used during the Second World War. A record kept by Bill Austin shows that, in the year ending 31th March 1940, precisely 4,279,141 lb of paper were converted into nitrocellulose for use in the manufacture of cordite.

During this time, paper was supplied by three manufacturers and this was blended in the scrolling machines. Disley's paper was always used at the corrugating end because it had been shown to be more readily crimped than that supplied from elsewhere. Either Turner's or Wycombe's paper was used at the plain end of the scrolling machines. The paper rolls were colour coded to assist in identification.

Dr Reginald Sidney Gilbert Knight came to the RNCF in 1915 from the Royal Gunpowder Factory at Waltham Abbey as chemist-in-charge of nitroglycerine and cordite, which comprised a combined section until 1931 when nitroglycerine became a separate section. Later he became Assistant Manager and finally he was appointed Manager and Deputy Superintendent (MDS) in 1931, following the retirement of Mr W T Thomson, OBE. It was in 1938 that Dr Knight was himself awarded the OBE in the New Year's Honours list. A bridge which crossed the No.2 narrow-gauge railway track (now Blackhill Road), as it passed through a cutting between the Nitroglycerine Hill area and the cordite drying stoves, was called Knights Bridge and this was recognised to the extent that the name appeared on the site plans. The reason for this was that the bridge was frequently used by Dr Knight as he walked between the building identified as 'X3' on the site plan, which was the nitroglycerine Cordite combined office, and the nitroglycerine production area.

Another feature named after an individual was to be found three hundred yards to the east along this cutting where there was a junction called Thompson's Corner, often thought to be a misspelling of Mr W T Thomson's name, but this was not so. The name came into common use during the construction of the factory in 1915 when the General Foreman of the Topham Jones & Railton organisation was a Mr Thompson who had his hut situated at this point.

Left: Paper drying machine loaded with paper rolls.

Above: Nitrocellulose factory: Early pattern nitration pans.

Right: Nitration pans fitted with fume extraction hoods.

Opposite: A paper cutting machine.

Right: Paper nitration: Ceramic nitration pans loaded with mixed acids preparatory to the addition of cellulosic material.

Left: Once loaded with mixed acids, fume hoods would be lowered over the pans and the paper or cotton introduced. An extraction system was employed to remove the worst of the fumes, but the atmosphere in the nitration house would remain rather unpleasant.

Above: An electric truck used to transport nitrated paper from the nitration pans for subsequent processing.

Right: The interior of the guncotton vat house. Here the guncotton was boiled by steam injection in a further purification process.

Above: An operator loading scrolled paper into Nathan & Thomson displacement pans – the earthenware vessels in which the paper was nitrated. This replaced the necessity of manual dipping.

Two views within the guncotton vat house. The boiling of the nitrated material in this plant was the first stage of purification which removed much of the residual acid.

Right: The interior of a steam-heated vat showing the heating coil and kier.

Left: Raw nitrocellulose being loaded into one of a series of vats similar to that shown above.

Right: Unloading nitrocellulose from the vats after processing.

Left and opposite: A further stage in the purification of nitrocellulose, where the material is blended for consistency and pulped to form slurry. These 400lb beating machines used in the final processing of nitrocellulose would be familiar to people employed in the paper making industry.

5

Right: These vessels, known as 'potchers' – equivalent to the same machine used in the paper making industry – represent a further stage in the purification of the nitrocellulose slurry.

Left: Filter banks and associated grit runs, another ancillary process in the purification of nitrocellulose.

Right: The nitrocellulose slurry was relieved of any remaining grit by high-speed centrifuges, seen here, before being passed to the filters illustrated on the previous page.

Left: Nitrocellulose pulping moulding and blending – pulp concentrators.

Right: These tanks held the purified and blended nitrocellulose slurry. This was then pumped as required (via further tanks in the intermediate house) to the mixing houses, to be combined with nitroglycerine for the first stage of cordite manufacture.

Left: An overhead view of a group of centrifuges, two of which are loaded with nitrocellulose.

Right: Nitrocellulose pulping, moulding and blending – loading dried nitrocellulose into a breaking machine.

Below right: A typical service guncotton press.

Below: Unloading a nitrocellulose blending drum.

Above: Guncotton could be used alone as an explosive in certain cases. It would seem that Holton Heath, in earlier times, was able to supply these charges shaped and formed as required – hence the moulding and pressing of the material as shown here. The clothing worn by the workers in these pictures would date them as probably pre-WW2.

Above: A service guncotton drill in use during the First World War.

Chapter 11

THE BATCH-WISE PRODUCTION OF NITROGLYCERINE

Nitroglycerine

A site plan dated 1922 shows that nitroglycerine production involved the use of a charge house, a nitrator-separator, a washing house and a wash-water settling house. The RNCF nitroglycerine plant used Nathan, Thomson and Rintoul nitrator-separators which were so named in recognition of the innovative work on the nitration of glycerine carried out at the Royal Gunpowder Factory in the early years of the century. At this time, Lt Col Sir F L Nathan was the superintendent, Mr J M Thomson was the manager and Mr W Rintoul was in charge of the plant.

A diagram of this new improved nitrator-separator is shown overleaf. The unit consisted of a cylindrical lead vessel 'a', with the bottom sloping in one direction and containing lead cooling coils 'h' and air pipes 'g'. The small cylindrical portion 'e', at the top of the conical cover, was fitted with inspection windows 'f'. The only other fitting in the cover was a gland to accept a thermometer, 's'. A pipe 'k', left one side of the cylinder immediately after which it divided. The vertical tube 'm', provided a means for the removal of fumes by suction, while the sloping pipe 'k', opened out into an inclined gutter which lead directly to the pre-wash tank. Mixed acid was introduced into the apparatus via pipes 'd', which entered at the lowest point. This was so arranged in order that nitroglycerine could not get into the mixed acid lines. There were two branches to this pipe, 'b' and 'c', each fitted with an earthenware stopcock. One led to the waste acid 'egg', while the second led to the drowning tank.

Plant of this kind is described by Marshall in the second edition of his well-known work on explosives which appeared in 1917 and at that time it was the last word on nitroglycerine production. In the absence of exact knowledge of details of the RNCF process, reference has been made to Marshall to fill in the gaps.

The more general schematic arrangement showing the two nitrator-separators is also shown together with a diagram showing a complete plant, which is very similar to that installed at RNCF. The major differences were that the ABC house contained nitrator-separator, pre-wash and final wash tanks.

The plant installed in the AB house more closely approached the scheme depicted in the diagram opposite because, in this case, the final washing was carried out in a separate building.

Below: This photograph, taken in 1917, shows the inside of the ABC house in the nitroglycerine factory. The twin nitrators, separator and pre-wash tanks can be seen. Note the rubber tubes through which the nitroglycerine ran from one lead vessel to the next. Pictured at right is Mr Hollingberry, Principal Foreman of nitroglycerine, and at left Mr Norton who was to lose his life in the explosion of 1931. This picture must have been regarded as a classic illustration of the time as it was reproduced in many publications relating to explosives.

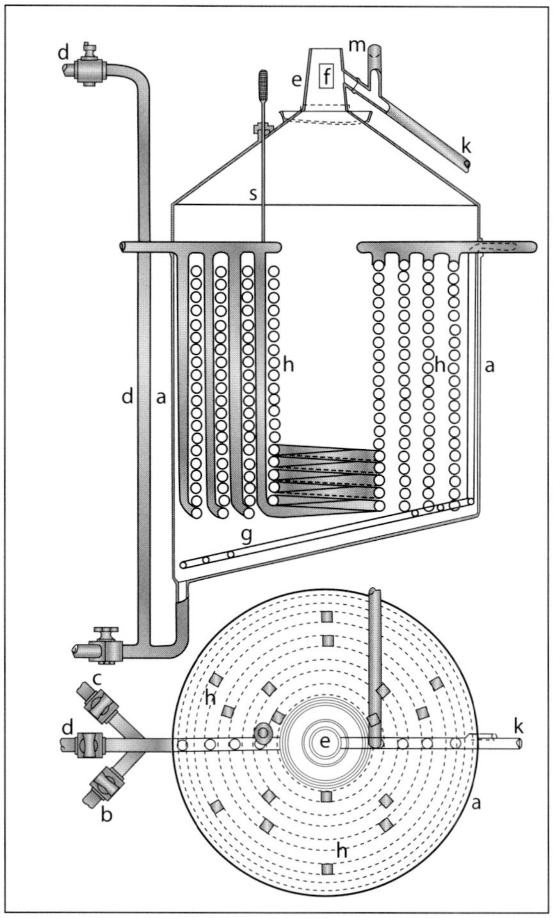

Above: Nitrator-separator used in the batchwise production of nitroglycerine.

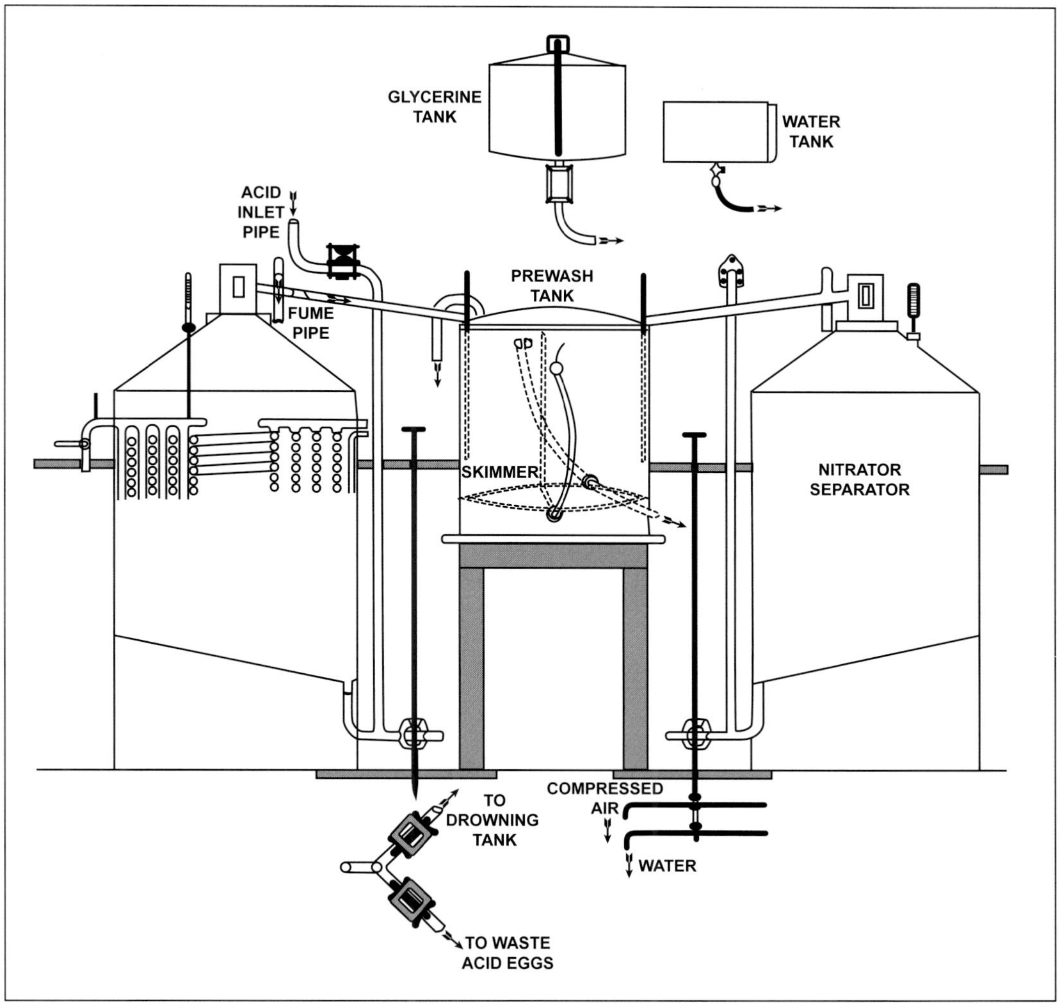

Above right: A pair of nitrator-separators and pre-wash tank of the type installed at RNCF.

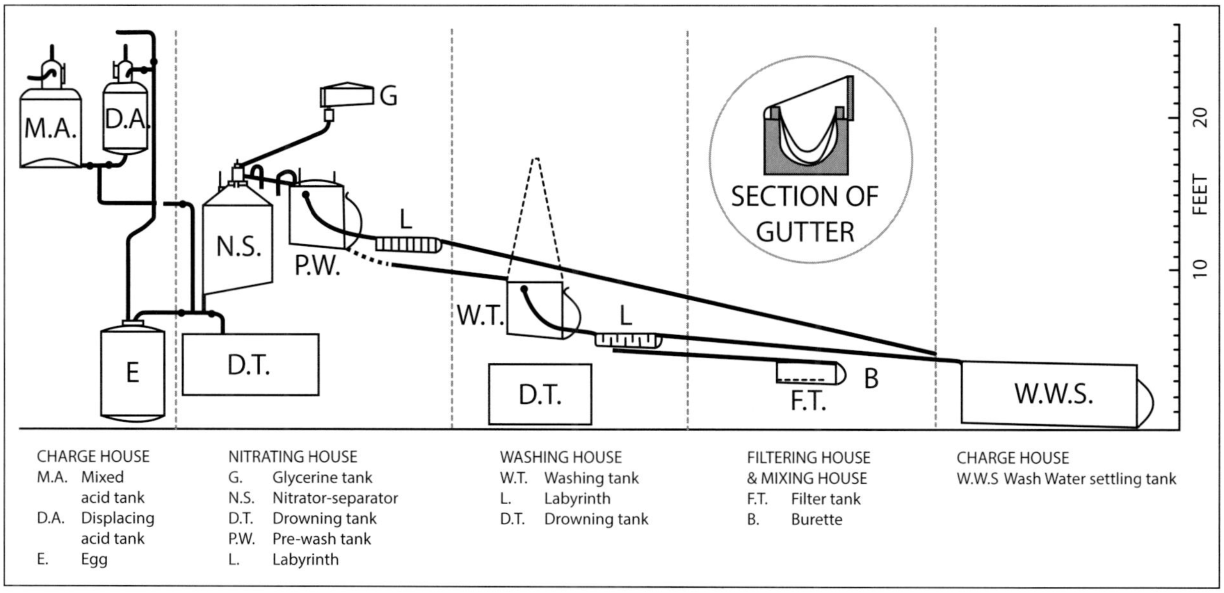

Right: A schematic illustration of a batchwise nitroglycerine plant of the type used at RNCF.

Above: View from Nitroglycerine Hill showing drying stoves in the far distance.

To avoid carrying or pumping the liquid explosive, it was conveyed from one process to the next by allowing it to flow gently under the influence of gravity along lead-lined gutters. All the reaction vessels, cooling coils and associated hardware were also made of lead which is not only acid resistant but also so soft that sparks could not be struck from it.

At this time, materials such as stainless steel were not available and the plastics industry which we depend upon so greatly today was in its infancy. Nitroglycerine was generally transferred from one vessel to another through soft rubber tubing which was pushed over lead outlet pipes and secured in place with oiled leather thongs. Control of the flow was achieved by kinking the rubber tube and in this way the potential hazard of using stop-cocks – which might initiate an explosion by friction – was avoided. The quality and size of all rubber in use at the RNCF was very tightly specified and must have caused the suppliers of this naturally occurring material quite a problem with compliance.

The floors of production houses at RNCF were covered with lead, making a continuous lining which at the walls and at entrances was dished so that nitroglycerine could be contained in the event of spillage. Artificial lighting was provided by electric lamps situated outside the building and arranged to shine in through the windows. All wiring to these was contained in metal conduits which were also run outside the building.

The workers had to make a complete change of clothing before starting work. The uniform was of a very hard, highly twisted cotton or woollen worsted cloth called 'lasting' and was black in colour until 1941 when wartime conditions made black material scarce and an equivalent white material was used. The fabric was very durable and also closely woven so it no doubt offered some limited protection in the event of fire.

This material was used for hats, coats and trousers. The uniforms were unusual in that they had neither buttons nor pockets. Instead of using buttons, the clothes were held together by tapes but there were no pockets and even handkerchiefs were supplied, the object being to prevent workers carrying personal belongings into the danger area. There were searchers employed to prevent unauthorised articles, particularly matches, being taken into the explosives production area. To complete the picture, the well-dressed Nitroglycerine Hillman also wore rubber half-wellingtons known as half-deckers, or just deckers, which offered a measure of protection against acid.

In RNCF practice, the batch-wise production of nitroglycerine commenced with the introduction of 8,1301b of mixed acid into the nitrator. This acid mixture, made up of 45 parts by weight of nitric acid to 55 parts by weight of sulphuric acid, was stirred by injection of air and cooled to the preferred nitration temperature of 10°C by the passage of chilled brine through the lead coils within the nitrating vessel.

This cooling is always necessary when glycerine is nitrated since the chemical reaction is exothermic, that is to say it produces heat. If the heat is not removed, the temperature rise could be such that the reaction could become uncontrolled and the nitroglycerine would catch fire or explode.

The glycerine charge was varied slightly to obtain the best yield but it was normally about 1,475lb and this was maintained at about 30°C when its viscosity is such that it could be readily introduced into the nitrator on the surface of the continually stirred and cooled acid. The glycerine injector was a separate part of the apparatus which was hoisted by means of a rope into the nitrating vessel when required. The glycerine was nitrated very rapidly under these conditions and much heat was generated. The temperature of the liquid contents of the reaction vessel was kept at 10°C by careful adjustment of the air stirring, brine cooling and rate of glycerine spraying.

The nitration of this quantity of glycerine, which yielded about one and a half tons of nitroglycerine, took about sixty minutes on average. This time could vary considerably however and it is certainly easy to imagine a cautious man taking quite a bit longer. Some operators on the other hand had a reputation for completing this process very quickly. Dr Ramsay in his notes of irregularities on nitroglycerine plant refers, for example, to a Mr Webster who 'was a notoriously fast nitrator'.

Alec Webster was an interesting character and there were many anecdotes about him. He was usually known as Tom, after the well-known Tom Webster who was for many years the sporting cartoonist of the *Daily Mail* and invented the character known as the Horizontal Heavy-weight. It was 'Tom' Webster, for example, who saw the factory chauffeur, Reg Cake, emerge from the Fleetsbridge Hotel and who subsequently reported the incident to the Superintendent. The story goes that Captain Morgan, the Superintendent at the time, said that if he had waited a little longer, he would have seen him (Morgan) come out as well.

The optimum temperature for nitration was 10°C, but occasional excursions to slightly higher temperatures were inevitable although whatever happened, the danger level of 22°C was never to be exceeded. If the temperature reached this level, the reaction was deemed to be out of control and the entire contents of the nitrator had to be 'drowned' in the large lead-lined, water-filled tank provided for the purpose directly underneath the gallery which supported the nitrators.

All that was required was to rotate a large stop-cock through a quarter of a turn. If this had become necessary, the operators were to make good their escape as quickly as possible, either via a bridge or through an access tunnel, depending upon their position at the time. It is understood that this procedure was never carried out at Holton Heath other than for test purposes, and inadvertently on a later occasion in AB2, due to instrument failure.

At the conclusion of a successful nitration, the air was shut off and the mixture was allowed to stand for one hour during which time the nitroglycerine separated out. The next stage, that of separation, was then undertaken. The lower density nitroglycerine, which had separated into an oily layer on the surface of the acid was displaced by the slow introduction of more waste acid from a former charge into the bottom of the nitrator. The product then overflowed the nitrator and ran down a short gutter into the pre-wash tank. The flow was watched through the sight window and was stopped before any acid passed over.

The acid nitroglycerine was displaced in three portions and each portion was pre-washed with an excess of water by agitation with air, which was injected into the tank in much the same way as in the nitrator-separator. The air was then turned off and the water-immiscible product was allowed to settle at the bottom of the tank. The water was then skimmed off the nitroglycerine by means of a funnel and a rubber pipe device and conducted into a so-called 'labyrinth', a vessel designed as a trap for any remaining explosive. From this, the washings proceeded to the wash-water settling house, along a lead-lined gutter, where any final traces of nitroglycerine were allowed to settle out for recovery. This washing process was

Below: View from Reservoir Hill over AB1 showing cooling towers and Poole harbour in the far distance.

Above: Mixed acid storage tanks for nitroglycerine.

repeated once more with water and once with a dilute solution of washing soda.

The product then received a further series of about five washings in another lead tank very similar to the pre-wash tank. In the case of nitroglycerine prepared in the ABC house – which was later renamed AB1 – this operation took place in the same building but material from the AB house was carried along a lead-lined gutter to a separate washing house C2, before the final washing could commence. The first of these washings was with dilute washing soda solution and the last few were with warm water at about 30°C. The product was finally run through a sponge filter and provided that it then passed a special heat test, it was ready for use.

The first batch of nitroglycerine came from the ABC house on Thursday, 12th October 1916 and the last was produced on Tuesday, 18th September 1945. The date of the first production batch from the AB house is not known but it could not have been much earlier because the first batch of guncotton was not produced until Monday, 7th August 1916 and this had to be dried before it could be used to make cordite. The last batch was made in this house on Tuesday, 23rd June 1931; the day of the great explosion.

Before a batch of nitroglycerine could be sent down the gutter to a 'C' house for washing, some communication between the workers in the two buildings was obviously necessary and this was accomplished by means of a simple signalling system. The signalling arrangement involved the use of back-lit stencils. Stencils bearing the various standard messages were duplicated in each building where they were mounted in window frames and illuminated by throwing appropriate switches, which energised electric lamps mounted outside the building so that light shone through the stencils. AB1, for example, could signal 'are you ready' and the appropriate stencil would be illuminated in both houses. 'C' house could signal back 'not ready' or 'send charge'. When the nitroglycerine had been sent down, the standard practice was to send a measure of soda after it and when the last portion had been sent, the gutter was wiped down, starting at the AB1 end and finishing at the washing house. As soon as the soda had been sent, then AB1 posted the signal 'charge down'. Finally, once the charge had been received, at the washing house, workers there switched off the 'send charge' signal. A red light was also provided at either end and this was to be used in the event of an emergency.

It was as a result of a short cut in the wiping down procedure that a potentially dangerous situation arose in early February 1938. The principal foreman was making a routine Saturday morning inspection of the gutters when he found several large patches of frozen nitroglycerine on the wash water gutter cover (which was immediately under the nitroglycerine gutter), on the catwalk, on the woodwork and on the tarmac near the tunnel leading from AB1 to C1. An examination of the canvas cover on the nitroglycerine gutter showed that this had sagged into the gutter and deflected the liquid product over the side. A little detective work revealed that the two Nitroglycerine Hillmen who had been working in AB1 at the time had untied securing tapes prematurely in order to expedite the later wiping down operation.

Extensive contamination by nitroglycerine of the tarmac beneath was found and this was thawed with hot water. The tarmac was then forced up and broken by hand with drops and pools of explosive being removed as they appeared. The operation was hazardous because of the presence of grit mixed with the nitroglycerine. Frozen nitroglycerine is, in any case, very sensitive to shock. The upper layer of earth was also scooped up and this together with the tarmac was later dumped at sea. All contaminated woodwork and all exposed earth, together with the surrounding area, was liberally treated with methylated caustic soda to 'neutralise' the nitroglycerine. This treatment was repeated a few days later and the damaged tarmac was made good. About 121b of nitroglycerine was recovered. As a result of this affair, the two Nitroglycerine Hillmen were transferred from the section and a number of additional measures were introduced in order to prevent the possibility of a recurrence.

As a result of the Nitroglycerine Hill disaster of 1931, when a mixing house was involved in an explosion, the gutter method of transport between washing house and mixing house was abandoned and a mechanical transport system was introduced. This involved the use of rubber-wheeled trolleys which carried the liquid explosive along grooves in a specially constructed concrete runway which was finished with an asphalt coating.

Laying the concrete and asphalt must have called for a high degree of workmanship, since the fit with the wheels had to be most accurate. In addition, the grooves must have been regularly cleaned out to remove falling leaves and soil. The trolleys carried tinned copper tanks which were rubber-mounted on the chassis of the trolley. To transfer the charge into the truck, one end of the rubber inlet tube, attached at the other end to the closed container, was passed through the wall of the washing house and fastened to the base of the burette. The nitroglycerine then ran into the truck under the influence of gravity. A calculated amount of water was then run in to fill up any free space in the container. The presence of this water, together with metal baffles inside the container, effectively prevented any 'sloshing' of the contents when the truck was moved.

The truck was then propelled by hand along the runway to the burette house, above and to one side of the mixing house, where it was parked immediately adjacent to a helical mixer, commonly known as a tun dish. The nitroglycerine was then siphoned from the truck and passed through a rose into the tun dish where it mixed with a slurry of nitrocellulose. The siphon was started by means of back pressure generated by passing warm water into the container. The Nitroglycerine Hillman displaced the last drops from the truck by blowing through a mouthpiece which, when not in use, was kept nearby in a bath of sterilising liquid. In this way, the passing on of colds and other infections was reduced – though the possibilities of blowing one's head off remained unchanged.

At this stage, the two violently explosive materials, nitroglycerine and nitrocellulose, were brought together and the process of cordite manufacture could start.

ABC House

Right: An external view of nitroglycerine charge house ABC showing the acid overflow tank.

Left: Nitroglycerine production soda solution tanks.

Right: Mixed acid and waste acid tanks and control valves associated with charge house ABC.

Below: This photograph illustrates the construction of the gantry supporting pipelines in the vicinity of charge house ABC.

AB1

Right: The nitration house AB1. This is an early picture, since the earth bank at rear was later raised obscuring the view of the charge house. The charge house, although a simple wooden building and one of the oldest on site, stood without maintenance for many years until finally brought down in the storms of 1988.

Left: An internal view of one of the batch-wise nitroglycerine plants, believed to be the AB house destroyed in the 1931 explosion. The two handwheels of the drowning cocks are visible in the centre. The handwheels of the brine control valves are at the sides, one being partially obscured. The sight-glasses and short gutters along which the nitroglycerine runs into the pre-wash tank can also be seen. Note the large earthenware stop-cock which was used to control the flow of mixed acids into the nitrator.

Above: An external view of nitroglycerine nitrating house AB1.

Above right: A view of the nitration vessels in AB1.

Right: The interior of a Nathan-Thomson-Rintoul type of nitration vessel, showing the arrangement of coils through which cold brine was circulated, to maintain a safe reaction temperature. A further set of coils were perforated to allow air to bubble through, thus constantly mixing the contents. The materials of both coils and the tank, was lead. Similar vessels were used in the subsequent washing process carried out in the 'C' houses, although cooling was at that stage not required.

Above: Nitrating house AB1 process control panel.

Right: This view shows the emergency drowning tank in the lower level of nitrating house AB1.

C1 Wash House

Right: C1 nitroglycerine washing house. The asphalt tramway connects through the tunnel at the rear of the picture. At the far left, a section of the canvas-covered gutters may be seen, down which the nitroglycerine ran from the AB house.

Left: Nitroglycerine C1 washing house tanks and fume pots.

C3 Wash House

Right: An external view of nitroglycerine wash house 'C3'. Note the asphalt tramway to the lower left foreground.

Left: Nitroglycerine washing tanks in wash house C3.

Above right: Nitroglycerine filters and measuring vessel in wash house C3.

Left: Washing tanks and associated fume pipes in wash house C3

Right: Wash house C3 fume towers, located some distance from the main building.

'D' Houses

Right: The interior of a type 'D' wash water settling house showing the settling tank and labyrinth.

Opposite: Spraying a measured quantity of nitroglycerine onto guncotton in the first stage of the production of cordite MD.

Left: An overhead view of a 'D' house mud filter labyrinth.

Miscellaneous

Right: A hand-propelled nitroglycerine transporter. Note the composition-lined guides in the concrete runway.

Left: The interior of one of the compressor houses built to provide compressed air for use in the nitroglycerine production process.

Right and below: Two views of the Linde ammonia compressors which were the main components of the refrigeration plant that provided low-temperature brine to cool the nitroglycerine nitration vessels. The compressors, in conjunction with evaporators and condensers, worked on the same principle as a domestic refrigerator but on a much larger scale.

Chapter 12

THE SCHMID CONTINUOUS PROCESS

Following the tragic explosion of 1931, advantage was taken of the most modern technology and a Schmid continuous production plant was installed as a replacement for the destroyed batch-wise production unit formerly operating on the site. Production from the new plant commenced on Tuesday, 16 February 1937.

This process differed from the Nathan-Thomson-Rintoul batch-wise production method in a number of technical respects, but basically the process still depended upon the nitration of glycerine and all the traditional processes including separating and washing were still involved. It is true that being a newer process, more modern materials were used in the construction of the plant, with stainless steel and glass being extensively used. The plant was housed within the AB2 hill. The major components of the plant, the nitrator, main and intermediate separators were mounted in the upper storey of the central wooden building, itself surrounded by concrete and a hill of earth. The upper storey of the wooden building was at a level with the two entrances. A concrete column was included within this building, which rose from the base of the construction to about 4 feet above the upper storey floor and to this the nitrator was securely bolted. Above and close to the nitrator, was the waste acid displacement tank and to the right was grouped a number of controls and instruments for setting air pressures and rates of flow and for measuring temperatures.

All movement of liquid into the plant from the external charge house was initiated by air pressure and the ratio of acid/glycerine was, once calibrated, automatically maintained by a special 'ARCA' regulator. The three washing columns were tall glass tubes about 5 inches in diameter. These were mounted from the base in the lower storey where the air for agitation was injected. They projected through the floor of the upper storey to a height of about four feet, ending in a wide tube coupled up to the fume extraction system.

The system for fume removal involved considerable engineering in itself. Firstly, the fume outlets passed out of the wooden building into the small passage which separated the wooden building from the glazed brick-lined concrete surround. In one corner was mounted two condensate traps, one for acid and one for nitroglycerine. These could be emptied by simply lowering a rubber tube. The fume outlets from these traps then combined to a four inch pipe which led out to the fume tower outside of the hill. The vacuum which sucked out the fumes was caused by the venturi action of injecting compressed air into the fume tower and was, apparently, a particularly noisy process.

Further traps were included within the fume tower, including a water cascade system for recovery of some weak acid and to prevent condensing acid from raining down on passers-by. The lower storey of AB2 contained further washing apparatus and a good deal of plumbing – all in colour coded piping – but the greatest amount of room was taken up with the drowning tank into which the contents of the primary separator or the nitrator could be dumped if necessary. The instrumentation fitted to the nitrator and separator included temperature gauges fitted with electrical contacts which would initiate an automatic drowning sequence should temperature rise above predetermined limits. The initial sensors as fitted were found to be rather unreliable and were responsible for a number of unnecessary drownings.

The drowning gear itself was a very complex apparatus, since it not only drowned the charge but – by today's standards – rather crude mechanical methods were used to ensure that the acid and glycerine feeds were shut down in the correct sequence. A Schmid plant was destroyed at Ardeer in 1940 when the glycerine feed was not stopped by this mechanism, as it should have been, during a power failure. The resulting temperature rise was such that the nitrator caught fire. At Holton Heath, the motive power for the drowning apparatus was provided by a falling weight and the delay apparatus for the acid shut-off was housed in an annexe alongside one of the entrances.

The upper storey of AB2 also contained gauges and controls for adjusting and reading the temperature and flow rates within the washing columns and the agitation air pressure. The indicators were logically grouped together on panels next to the appropriate washers. A small hand-washing sink was provided to allow washing, in water or soda, of residual amounts of nitroglycerine which were left at the end of shift.

Above: This picture, from 1946, shows the Nitroglycerine Hillmen and section officers of the time. As the manufacture of nitroglycerine at RNCF was shortly to cease, these gentlemen were probably the last Hillmen of Holton Heath.
Left to Right (standing): Harry Gover, Tom Hillier, Wilfred Fripp, Len Thrasher, Arthur Marsh, Herbert Eckton, Jack Vincent and Bert Mitchell.
(seated): Reg. Reeves, Fred Masters, Tom Williams (chemist) and Charlie Stockley.

A potentially dangerous situation arose on one occasion when some impure nitroglycerine was left in the sink and forgotten. When the next chemist came on shift this material was observed to be in an unstable state and giving off red fumes. Instead of running for his life, as the chemist would have been justified in doing, he calmly sent the material into the drowning tank, probably with very little time to spare, and thus prevented a dangerous accident.

The manufacturing process was commenced by mixed acid being injected into the bottom of the nitrator and glycerine injected into the top in the ratio of 5.5:1 of acid to glycerine. Cold brine was circulated throughout the process to keep the reaction under control. This was controlled as necessary by a large hand-wheel to the left of the nitrator. The contents of the nitrator were stirred in order to ensure uniform cooling, the stirring mechanism being activated by a gearbox which conveyed power from an overhead motor.

Nitroglycerine was produced at the rate of about 1,320lb per hour. An unstable emulsion of nitroglycerine in acid overflowed the top of the nitrator and ran into the separator, already filled with waste acid, where it separated into two layers with nitroglycerine at the top. The waste acid ran off from the bottom at a rate controlled by

Left: The Schmid nitrator inside AB2. The large vessel to the left is the nitrator, with acid and glycerine entering from right. The large hand-wheel to the left controls the flow of cooling brine. Instruments are provided to show pressure, flow-rate and temperature.

Above: This unusual looking equipment is the primary separator for the Schmid nitrator in AB2. At the rear may be seen one of the three glass washing columns which protruded into the room below.

a cock while the product again overflowed the top. The waste acid, before leaving the plant for reprocessing, was diluted with water in order to render any remaining nitroglycerine soluble and thus less dangerous.

The crude product was then injected with mains water into the bottom of No.1 washing column. Low pressure air was injected to ensure thorough agitation in the washing process and in this way much acid contamination was removed. The explosive and water mixture then overflowed into an intermediate separator and here the product fell to the bottom while the contaminated water overflowed the top to the waste water system.

The partially refined product was then passed through Nos.2 and 3 columns, each with its own intermediate separator, using warm water containing 3½ percent washing soda. The nitroglycerine was then injected together with warm softened water into the transport tube in which it travelled as an emulsion to the final washing house. The rubber transport tubes were jacketed with lead-lined gutters carrying warm water at 50°C as they passed out of the building.

It was in 1927 that the discovery was made that a nitroglycerine and water emulsion is safe within certain limits provided that water is the continuous phase. The transport tubes were fitted with a detonation break at about the half-way stage, this being a change in diameter of the pipe which led to the re-establishment of turbulence in the flow. As a result, water became once again the continuous phase and the emulsion was stable and non-explosive.

At the washing house, the emulsion was received into another intermediate separator after which it was given a final wash. The nitroglycerine and water mixture ran from this column into a lead hand-wash tank where it was allowed to separate.

The liquid explosive was then run under gravity into a large rectangular lead tank fitted with filter pads which removed water globules and any solid matter remaining. After successfully passing stability tests known as 'heat tests', the product was ready for use and it was measured into a lead burette and trucked away for use in just the same way as nitroglycerine produced by the batch process.

Nitroglycerine has a rather curious property that even now is

Left: This photograph encapsulates a number of interesting points. The view is from the site centre, looking towards the A351. The footpath to the right is now the centre of Blackhill road. To the left is the AB2 hill and the timber building is the charge house which supplied it. To the right of and behind the charge house can be seen a part of Knights Bridge where it crossed the railway track. The mounded area just behind the compressed air tank houses the waste acid recovery tanks.

The wooden steps leading up the hill provided access to certain air pipes which were elevated to prevent a siphoning action in the event of an air failure causing nitroglycerine to be drawn back into the air system.

not fully understood. This involves the spontaneous firing – with a green flame – of textile material in the presence of acid and air. The proportions of this combination appear to be critical and it is not easy to recreate the effect in the laboratory. When it occurs in practice it can have devastating effects and two quite serious accidents attributed to this phenomenon occurred during the Second World War at Bishopton and Wrexham. This phenomenon manifested itself at least twice at RNCF but fortunately there was no harm done on either occasion.

The first of these took place on a Sunday morning in 1942 when Bert Mitchell was checking acid lines under air pressure as part of a routine maintenance operation. The air pressure was released by turning a valve and when he did this, acid splashed out onto his nitroglycerine-contaminated lasting jacket which immediately and spontaneously fired with the characteristic green flame. Also present were Nitroglycerine Hillman Marsh, Assistant Foreman Reggie Reeves and Chemist John Gray who was supervising the operation. Arthur Marsh turned a 1-inch hose on to the flames which were soon put out, but John Gray will carry to the grave the picture of utter astonishment on Bert Mitchell's face as he beat at the flames with his hat!

The second example occurred at the AB1 plant. On this occasion, Bert Fricker was withdrawing the glycerine injector after a nitration in the usual way with a block and tackle when the rope broke. Bert caught the injector on its way down and his jacket fired. The chemist who was present threw a bucket of water over him and again no harm was done. The rope broke as a result of acid fumes which had rapidly degraded its strength. This rope failure occurred even though the effect was well known and all ropes were frequently examined and changed. The use of chains was precluded because the possibility of metallic friction in the presence of nitroglycerine was an unacceptable hazard.

Because of this little-understood phenomenon, in nitrating houses, where acid and nitroglycerine are both present, separate cleaning cloths were used for wiping up traces of these materials. When not in use they were kept in rubber buckets – under water.

Left: This photograph from 1987 shows the interior of AB2 after de-commissioning. The lower half of the building has been filled with soil, burying the drowning tank. The concrete column in the foreground was the mounting post for the Schmid nitrator. To the rear may be seen one of the two access tunnels. All traces of the wooden interior have been removed.

Chapter 13

CORDITE PRODUCTION METHODS

Over the years, various forms of cordite were produced at RNCF. During the First World War a little of the original cordite MkI was produced but the bulk production was of cordite MD. This was preferred because it did not erode gun barrels as severely as did the MkI material. Both types were made in the same way by the classical solvent process and both contained the same materials – nitrocellulose, nitroglycerine and mineral jelly – but the ratio of nitrocellulose to nitroglycerine was much greater in the MD version.

In 1927 the solvent process was abandoned and solventless cordite, or cordite SC, was introduced. This involved major changes in manufacturing technique and the composition of the cordite itself was changed with the ratio of nitrocellulose to nitroglycerine being between that used in the MkI and MD materials. At the same time the nitrogen content of the nitrocellulose was also reduced. The mineral jelly was replaced by another stabiliser known as Carbamite or Centralite. This component also assisted in the gelatinisation of the cordite in the absence of acetone, the conventional solvent which had been used hitherto. Apart from certain disadvantages associated with much greater stiffness at the pressing stage, it had many advantages in terms of safety. It avoided, for example, the well known hazards associated with manual handling of dry guncotton and the nitroglycerine and guncotton paste and it also eliminated the need for highly flammable solvents. It was also a more efficient manufacturing method because the long and expensive stoving processes of both guncotton and cordite were avoided. In addition, the SC cordite was more stable, less prone to irregular shrinkage – which was a common feature of the earlier cordites – and it also burned with a cooler flame.

Below: Cordite MD – incorporating and pressing ranges.

One of the major problems with propellants of the SC type was that they gave rise to an enormous flash which gave away the position of the gun, the flash being the secondary ignition of gases released with the shell. Efforts were made to find means of eliminating or reducing the flash and the most effective way was found to be to introduce into the cordite another ingredient which reduced the temperature of combustion.

One of the most satisfactory materials for this purpose was picrite, or nitro-guanidine, and this additive was introduced into Naval service in cordites of the NF type in 1939. However, because picrite is soluble in water, it was found necessary to return again to the use of solvents in the manufacturing process. Further details of picrite manufacture will be given later.

No matter which form of cordite was being made the essence of the process was the bringing together of those two most formidable materials, nitrocellulose and nitroglycerine and to finish up with the solid, stable product in the desired physical form.

Cordites MkI and MD

The traditional solvent process involved, first of all, the drying of the nitrocellulose. This process has already been described in the section dealing with nitrocellulose, which for cordite MD was, in fact, guncotton, because of its high nitrogen content. (Guncotton. it will be remembered is the name given in the armed services to nitrocellulose containing 13% or more of nitrogen.)

Dry guncotton was then weighed into waterproof, rubberised bags which were taken to the paste mixing houses where a measured quantity of nitroglycerine was poured onto the guncotton. This mixture was emptied onto a lead table where the charge was then given a preliminary kneading by hand before being gently pressed

Below: An official picture of women cordite workers, taken close to the press houses. The photograph shows the improved protective clothing which was issued after a serious accident involving a cordite fire.

through a sieve of half-inch holes at the end of the table when it fell into a bag attached below. This must surely have been a perilous operation. Marshall suggests, superfluously one would have thought, that 'no unnecessary violence must be used.' The so-called 'paste' thus obtained was transferred next to the incorporating house where it was placed in an incorporator. These machines were rather like bakers' dough mixing machines, which could be warmed or cooled as appropriate with a water jacket. The correct amount of acetone was then added and the mixture was blended together for a period of about three hours, after which the mineral jelly was added to the extent of 5 percent of the total mix, after which the incorporation process was continued for a further period of about three hours.

At this stage the product, now called 'dough', was taken to the presses on the cordite ranges towards the north-west of the factory (in the corner now bounded by the A351 and Blackhill Road). Here the cordite dough was pressed through a die in a hydraulic press

Below: A 'fireless' locomotive outside a press house during the manufacture of cordite MD. The tall, thin wooden tower to the rear of the engine is a hydraulic accumulator consisting of a large weight which would be pumped up to the top. The potential energy thus stored would then be used as a smooth supply of hydraulic power, via a weight actuated piston, for the presses. A similar system was designed to operate Tower Bridge in London.

behind a rope mantlet screen, from which it emerged in the familiar cord-like form. The product was collected by hand as it emerged from the presses and cut into measured lengths.

The cords were loaded into wooden trays of the kind already described which were, in turn, placed in a truck and taken to the drying stoves. In the stove area the trays were placed initially in the acetone recovery stoves. Here the trays were stacked in chambers heated by steam pipes. As each chamber was filled it was sealed with the exception of a small opening at the bottom of the door. Air was drawn by suction through the stoves and over the cordite, which was maintained at a temperature of about 43°C.

The acetone vapour passed through a pipeline, supported about eightfeet from the ground, to the acetone recovery plant. After a period of about one week, when the acetone given off had diminished to such a low level that it became uneconomic to continue the recovery process, the cordite was removed and taken to the final drying stoves. These were similar to those just described except that no acetone recovery equipment was fitted.

This final drying period varied according to the diameter of the cords but it could take as long as four or five weeks. Finally, after drying, the cordite was blended in one or more of several ways to ensure a constant quality product with uniform characteristics. All that then remained was for it to be packed, bonded and despatched by rail or primarily, by barge from Rocklea Jetty to Priddy's Hard or to Upnor Castle on the River Medway in Kent.

Below: Cordite MD – incorporating and pressing ranges.

Above: Electric drive mechanism for cordite MD incorporating machine.

Right: Cordite MD incorporating machine in its operating position with cover in place. Note the drive-shaft, with flywheel, coming through the wall from the motor and gearbox assembly on the far side.

Left: A cordite MD incorporating machine in the elevated position and with the cover removed to show the incorporating screws.

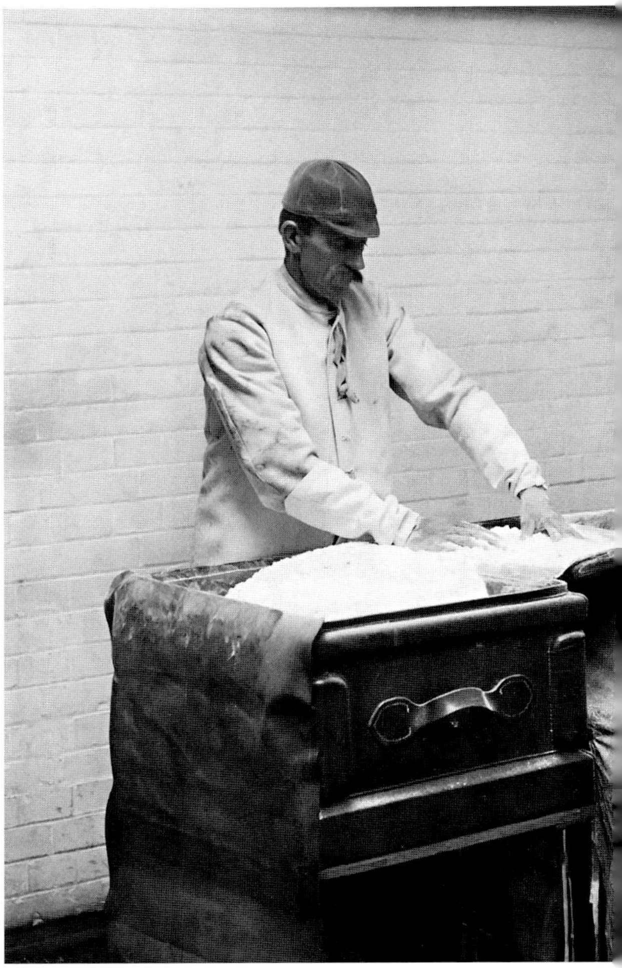

Above: Cordite MD – loading incorporating machine.

Left: Cordite MD – unloading incorporating machine.

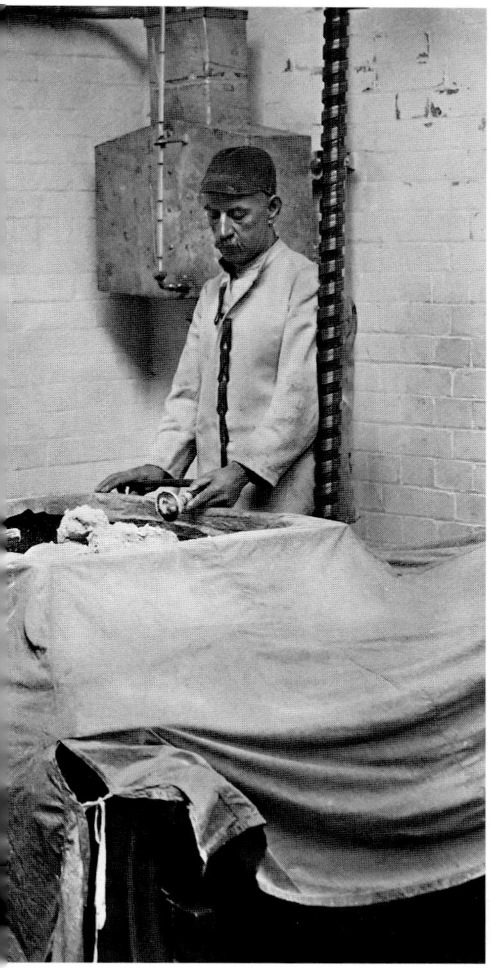

Right: Cordite MD hydraulic press. Note the heavy rope mantlets which could be drawn around the machine when in operation to give the staff some degree of protection against an accidental explosion.

Above: Cordite MD – Charging a hydraulic press cylinder in preparation for pressing.

Right: Cordite workers cutting the cords to length. The accident records show unwary fingers to have been trapped in the knife on more than one occasion.

Left: The interior of cordite drying stove S18. The cords were laid in the wooden trays at the right. Hot water was obtained from steam in a nearby calorifier, or heat exchanger, and circulated through the wall-pipes. This stove is typical of many in the factory.

Cordite SC

The stages in the manufacture of SC were briefly as follows. Firstly, the nitroglycerine and nitrocellulose were mixed under water in a tundish in the burette house. In the early days of SC production the nitrocellulose pulp was trucked to the paste sheeting and mixing houses, but since the introduction of the nitrocellulose ring main in 1932, as previously described, this was piped direct to the mixing houses mixed with water.

The aqueous mixture of nitrocellulose and nitroglycerine then flowed along a gutter into a mixing tank where it was kept agitated by motor-driven paddles. The carbamite gelatiniser/stabiliser and other minor additives were then added as slurry with water. During this time the nitrocellulose absorbed the other organic materials. The mixture was then considered safe to pump, albeit with a diaphragm pump with rubber valves, to the sheeting table where excess water was removed by draining under vacuum and pressing. Firm sheet of from ⅜ to ½ inch thickness were obtained which at this stage contained about 25 percent moisture.

Sheets weighing about 10lb were then loaded into large wooden trucks which were taken to the paste sheet drying houses where a current of air at 50°C could pass over the cordite paste. This process took about 24 hours.

The paste, still in its truck, was then taken to one of the rolling houses where the dry paste was gelatinised by passing it several times through heated rollers. The sheet cordite was then cut into 7 inch squares and placed in aluminium containers which held 251b. The rolling houses were fitted with chimneys over each rolling machine

Right: Cordite SC production - weighing the Carbamite stabiliser.

in order to ensure that heat from any fire which occurred would be directed upwards and out of the building. The forest of these large chimneys surmounted with metal extractors gave the rolling houses a very individual appearance.

The 251b charge was taken, still warm, to the charge forming houses where, after examination, it was consolidated in a small press to form what was known in the jargon as a 'cheese'. These cheeses were then taken in heated trucks to the special press houses where they were placed in the heated cylinders of presses.

The pressing operation was undoubtedly more hazardous with the SC material because of its extremely high viscosity. This meant that it had to be pressed at high temperature and pressure. Previously pressing was carried out behind rope mantlets with the workers in the same room but special precautions were gradually adopted for the pressing of SC cordite. In the light of experience gained as a result of several explosions in press houses, a new design was evolved for the layout of the buildings and gradually the sites of all the old guncotton drying stoves were occupied by these new structures. A much greater margin of safety was obtained by having three separate buildings – one for the operatives, one for the augmenters and motors and one for the cordite presses. Earth traverses were built around and between each press house. The presses were surrounded by steel framed rope mantlet curtains within a chamber of reinforced concrete. Because control of the pressing operation was conducted remotely from the press, workers were much better protected. The cutting room and connecting passageways were of light timber construction and visual communication was maintained by a system of mirrors.

Later still, in the 1940s, some very large presses were introduced for production of rocket propellant and catapult cordite which was used to assist the take-off of planes from aircraft carriers. These buildings were constructed on old press house sites and were of different construction again, being steel-framed and of reinforced concrete, with an interior of glazed brick and having the press rooms covered by a large earth mound which was subsequently turfed. P17 and P18 each housed two vertical presses which were approached along an isolating corridor having steel mantlet blast doors at the entrance to the press room. The presses sat in a sump some 4-5 feet deep. These buildings still exist in good condition although, due to the fact that the drains are blocked, they are normally slightly flooded. This makes entry for the unwary very hazardous, since it is very easy to stumble into the water filled sumps.

P14 and P15 housed horizontal presses. The press room was of reinforced concrete covered with earth but the cutting room, motor room and connecting passage were of timber which has long since disappeared. The largest press on the site was housed in P16 and the size of the underground white glazed brick lined press room, as could be seen in the mid-1990s, gives a fair impression of the immense machine it must have been. P16 is also in very good condition and, with its several large rooms, would be capable of housing a small business today, were it not for the fact that the Admiralty have filled in the old railway track cutting on that part of the site retained by them and as a consequence P16 is almost always flooded due to the lake that now forms to a depth of around fourteen inches.

A date on the overhead rolled steel joist, which was present in all the large press houses to assist in movement of heavy pieces of machinery, shows the last test to have been in 1952. All press houses were lit, as would be expected, either from outside or with intrinsically safe fittings. Some of the lighting fittings can still be seen.

Above: Adding Carbamite to paste during the cordite SC production process.

Left: Carbamite grinding mill.

Right: A pair of diaphragm pumps used to transfer carbamite-stabilised nitrocellulose / nitroglycerine mixture to the cordite SC sheeting tables.

Above: An interior view of a cordite SC mixing house showing a group of tanks and diaphragm pumps with a sheeting table in the background.

Right: Backwater tanks attached to a cordite SC mixing house.

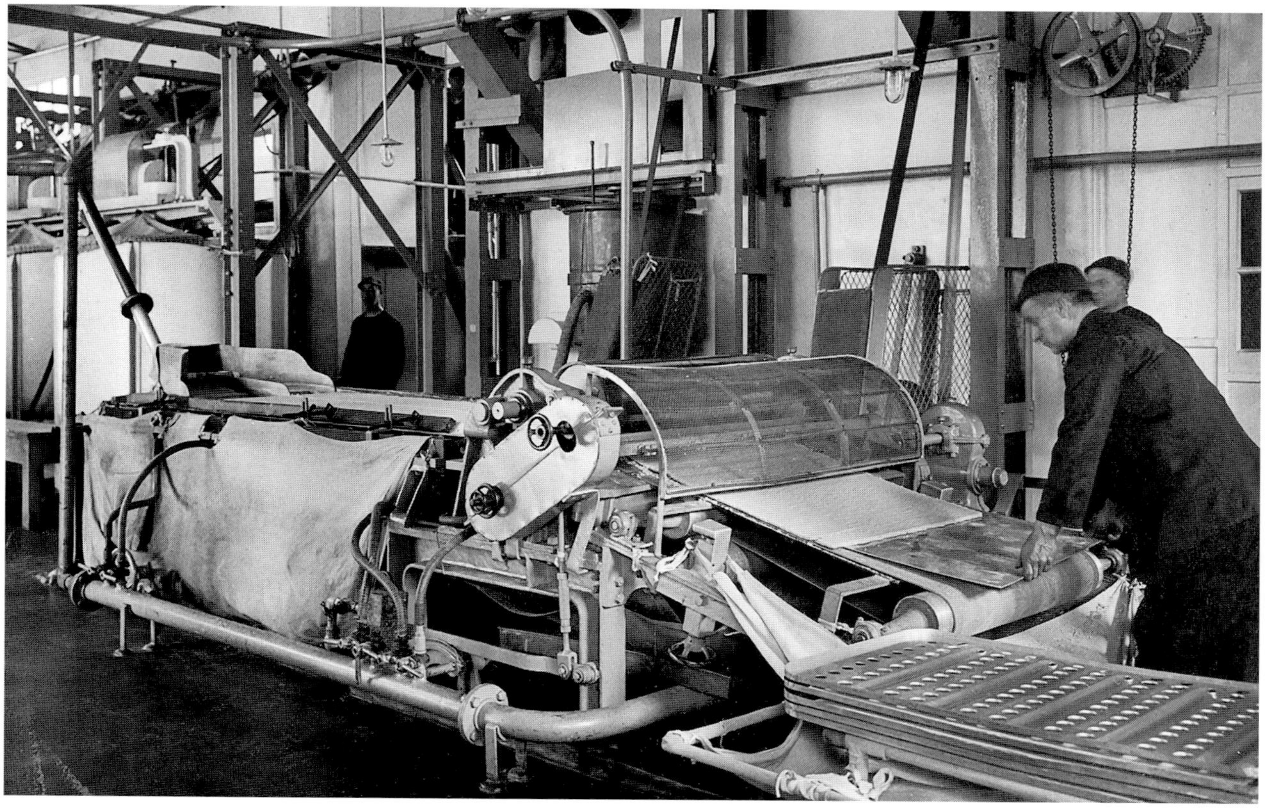

Left: Ben Davies collecting paste sheet in a Type 'E' mixing house. An identical machine was in use at the Royal Ordnance Factory, Bishopton.

Right: Cordite – SC paste drying air ducts for trucks.

Below: An exterior view of a cordite SC paste-drying air heating unit. The drying medium was hot air derived from a calorifier in the hut. This in turn was heated from the steam supply main.

Right: A wider view of the cordite SC paste drying arrangements showing, top centre, a calorifier hut from which the hot air ducts emerge and are attached to the inlet end of the drying trucks. To the right can be seen four large outlet ducts, two of which are connected to trucks by means of canvas attachments.

Below: An example of a type 'F' cordite SC sheet drying house and trucks.

Left: A detailed view of a cordite SC sheet drying truck showing the frames upon which the sheets are supported.

Below left: Interior view of a cordite SC charge forming house.

Right: A typical, large hydraulic cordite charge-forming press.

Below: Exterior view of a cordite SC charge forming house. The wooden tower adjacent is presumed to be a hydraulic accumulator.

Above: A view of one of the cordite rolling houses G3, taken on 6th March 1941. Note the forest of extractors on the roof which were intended to direct any fire upwards and away from the operator. These buildings have long since been replaced by modern industrial units. Note the guncotton vat house 'saw-tooth' roof visible at extreme right.

Opposite: Square cutting and packing cordite SC. Notice the open overhead line belting driving the cutting machine.

Above: A cordite rolling machine. Note the steam-heated rolls and the hood above leading to a roof chimney. The hood also contained a 'drench' of water as a further measure of protection.

Right: Press house P14. This house contained a large cordite press. The press room was underneath the earth mound. The wooden exterior structure housed the cutting room and connecting passage. The photograph dates from October 1937. Note the exterior lighting, obscured windows and lightning arrestors, all characteristics of a 'danger' building.

Left: Press house P18, shown here under construction during the 1940s, contained two large vertical cordite presses. The presses were located at the rear with an earth mound being constructed over.

Above: A vertical cordite press of the type installed in P17 and P18. This was remotely controlled.

Above right: This photograph shows the remote control and monitoring instrumentation as fitted in P16, P17 and P18. This enabled the operator to work in complete safety should a cordite fire occur.

Below right: A very large cordite press in P16, the largest press house on the site. This semi-underground building was still intact at the time of writing although flooded due to decay and blockage of the drainage system. It is believed that the horizontal presses were employed in making large diameter cordite sticks for use in aircraft assisted-take-off systems.

Above left: Detail view of the breech block mechanism of the cordite SC catapult charge forming press illustrated below.

Above right: Detailed view of the cordite sheet cutting machine illustrated on page 176.

Left: A general view of one of the horizontal catapult charge forming presses installed at Holton Heath in the 1940s.

Right: A detailed interior view of a cordite SC packing house.

Left: Another interior view of a cordite SC cutting and packing house. No the generous allocation of fire buckets, the stack of filled packing cases, and the special, pocketless safety clothing worn by the operatives.

Flashless Cordite.

Picrite was used in large quantities in flashless cordite but, since it is water soluble, its incorporation with the other major components required the re-adoption in some measure of the solvent-based process of cordite manufacture.

The first stage in the production of flashless cordite by this modern solvent process was paste mixing and this was carried out in the paste mixing houses in just the same way as for cordite SC. The resulting paste was then crumbled while still wet and then dried and delivered to the cordite section on trays. Because of the sensitive and highly flammable nature of the dry paste, special precautions had to be taken. The relative humidity was maintained above a certain level, for example, and the building was designed to prevent accumulation of dust. The weighing machines had their moving parts immersed in oil while water seals prevented the ingress of paste dust into the interior of these machines. All plant was earthed and operators wore special clothing.

Below: A cordite press in operation in the 1920s. One must assume that the large horizontal hand-wheel controls the flow of hydraulic fluid to the press, the pressure imposed by which is monitored by the gauge seen above the operator's head.

Weighed quantities of picrite were similarly despatched to the incorporator house in tared bags on trucks. Sodium cryolite was weighed into metal containers before it was delivered to the incorporation house. The carbamite, which was received as flakes, was ground at the carbamite grinding house before bags containing weighed quantities were delivered to the same incorporation house.

Having assembled all the ingredients in the appropriate quantities, a portable table was placed beside the incorporator machine and acetone was introduced to a measuring burette on the wall. The bags of paste and picrite were then brought to the table and emptied on

Below: Percy White (left foreground) and fellow workers, laying out flashless cordite. The presses were sited behind the heavy protective rope mantlet hanging in the background. This photograph was almost certainly taken on the Z ranges.

to it. The material was examined for the presence of any obvious foreign material and it was gently pushed into the machine. As this happened, acetone was sprayed onto the mixture of paste and picrite while the blades of the incorporator rotated.

When paste, picrite and acetone had been loaded, the lid was placed on the machine and it was allowed to run for half an hour. The lid was then removed and the carbamite and cryolite were added before incorporation was resumed. Gradually, the mixture changed from a powdery mass to a stiff dough as gelatinisation occurred. The whole process took about six hours and during this time hot or cold water was circulated to maintain the temperature at a suitable level and to ensure that it did not rise above a preset limit. At the end of the process, the incorporator was tilted and unloaded, the lumps of dough being placed in waterproof bags for transport to a press house.

Below: Press house P2 now derelict, shortly before removal of buildings during redevelopment of the site as the Admiralty Research Establishment.

Picrite

Picrite, the chemical name for which is nitroguanidine, was and still is a major component in flashless cordite. Picrite generates large volumes of nitrogen when the propellant burns and this has the effect of reducing barrel wear and muzzle flash. A decision was taken that flashless cordite (RDN) should be manufactured for naval use and it was further decided that this should be produced within the RNCF. The picrite factory was built over a period of about two years commencing in 1937. This involved designing and building twenty structures which included mess and changing rooms, lavatories and offices as well as all the buildings that comprised the production plant. The factory layout called for a central approach cutting with 2 feet 6-inch gauge railway tracks having access to all production buildings. The production buildings were all protected by surrounding earth traverses with some earth covering to the flat roofs. This gave a measure of protection from aerial attack, as well as offering a safety feature in the event of an internal explosion. Approach tunnels, including emergency tunnels, gave access to the buildings through the surrounding earthworks. The odd entrance tunnel is about all that remains of the picrite factory, as the buildings were broken up and covered over at the time the Blackhill Road was being made.

The buildings were mainly steel-framed, with brick exterior walls, concrete roofs, and lined inside with white glazed brick, with glass brick windows. Ventilation was achieved by use of extractor fans in the roof. The whole factory, including all the services, cost a little over £46,000 which really does seem a derisory sum for such a sophisticated and extensive installation.

The refrigeration houses built for the picrite factory are worthy of especial note because they possessed the unusual feature of having cooling ponds on their roofs. The refrigeration and compressor facilities were so important for picrite production that two identical buildings were constructed for the purpose. The following details refer to the first of these buildings, N1.

Because of the unstable nature of the soil it was necessary to resort to piling. No fewer than 109 reinforced concrete tubes were driven to a depth of about 36 feet. Once in place they were pumped out and reinforcing steel was lowered into place. Concrete was then poured. These served as the foundations of the building and the air and brine compressors. The roof was made of reinforced concrete with a built-

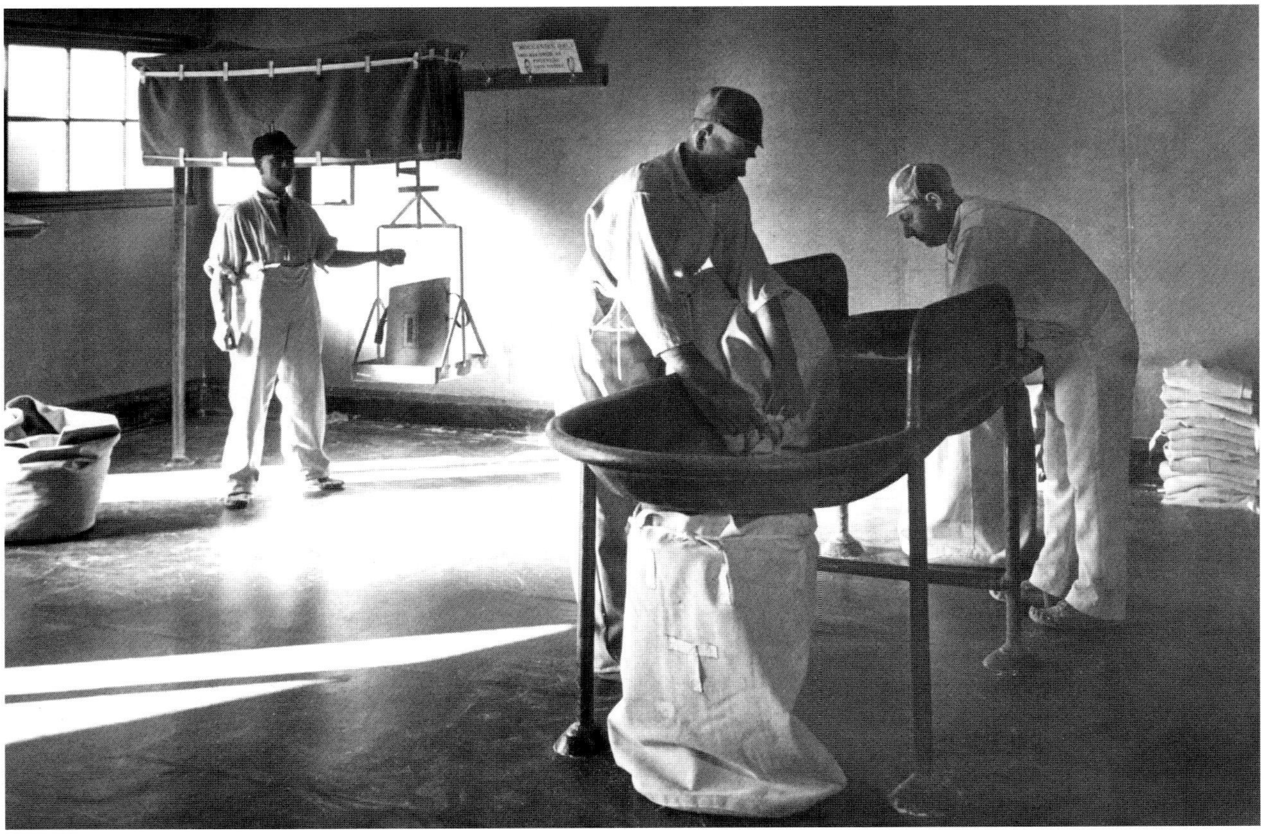

Left: Weighing dry paste and picrite and pre-mixing, prior to incorporation. Note the wearing of soft moccasins and the dust shield over the mechanism of the weighing machine. Dry paste is very inflammable and dusty. The is most likely to be within E6 mixing house.

Left: A picture of N11. one of two identical drying stoves for picrite. The trucks were stationed in the siding and coupled to the hot air ducting system. This method avoided the necessity of unloading and reloading the trucks.

up concrete kerb around the edges. This was lined with a bitumen-type membrane. The timber-slatted water-retaining side panels were supported by strutted steel framing on the concrete kerb. The design was very successful and was in use for six years without any leakage from the pond.

The production of picrite was a fairly simple process involving only two stages. Firstly, ammonium nitrate and dicyandiamide, known at RNCF as 'Product D', were fused together to give the chemical intermediate guanidine nitrate. Following the nitration process, which involved dehydrating the guanidine nitrate, nitro guanidine, or picrite, remained. In practice, the production was very much more complicated than this because the 'Product D' itself had first to be synthesised. Subsequently much attention had to be paid to the purification of the picrite and to the careful control of particle size in the finished product.

'Product D' was produced from calcium cyanamide, which was bought in, at the extraction house N6. Here the calcium cyanamide was hydrolysed by water to give cyanamide which, when melted, converted itself into dicyandiamide or 'Product D'. Lime was produced as a by-product and this was dumped at sea. Immediately prior to use, the 'Product D' and the co-reactant ammonium nitrate were thoroughly dried on aluminium trays in aluminium trucks at a drying stove (sometimes T6). The two reactants were dried in the same truck, but on different trays, for six hours at 80°C. The first stage in the main production of picrite then took place in fusion house N8. Here, the thoroughly dried co-reactants were intimately mixed together in the molecular proportions of one of 'Product D' to two of ammonium nitrate. When this mixture was heated in a fusion pot to about 180°C the two materials reacted together to give guanidine nitrate.

Standard batches of 40lb of 'Product D' and 84lb of ammonium nitrate were fused together at this stage and, because the reaction was an additive one and there were no by-products, 124lb of fused product was obtained. This chemical intermediate, guanidine nitrate, had then to be converted from a hot liquid into a convenient solid form before it could be used in the next process.

The exact method used at RNCF is not known but it is likely that the method used at the Royal Naval Propellant Factory Caerwent was modelled on the Holton Heath process and at Caerwent the molten guanidine nitrate was transferred to the surface of a rotating drum flaker. The water-cooled drum rotated and as the guanidine nitrate solidified on the surface it was removed in flake form by a stationary doctor blade.

The solid, flaked guanidine nitrate was then ready for the final stage in the chemical production process which took place in nitration house N9. The nitration process involved the interaction of 90% sulphuric acid with the guanidine nitrate. In the RNCF process 8501b of acid was added to 496lb of guanidine nitrate, the product of four fusions in N8, in a lead-lined cast iron nitrator fitted with brine cooling and air agitation. The acid, in the space of a few minutes, then performed its allotted task of removing the elements of water from the guanidine nitrate leaving crude picrite.

The powerful affinity of concentrated sulphuric acid for water is widely exploited in synthetic organic chemistry and at RNCF it was used in the manufacture of nitrocellulose, nitroglycerine, picrite, and, as we shall see later, of tetryl.

The nitrated charge was then 'drowned' by running it into a vat containing what is rather cosily known in chemists' jargon as 'mother liquor' and which was in this case the spent acid left from a previous nitration. This drowned charge was then centrifuged and about 3001b of crude picrite was obtained.

The next stage was concerned with refining the crude product and took place in the dissolving and crystallising house N10. Here the picrite was dissolved in hot water and passed through a filter bag while still hot to remove insoluble impurities. Pure picrite was then obtained by crystallisation from cold solution using a spray method.

Finally, the picrite was dried in wooden trucks by passing an air current at 50°C over the product at a drying stove, two of which are shown on the site plan as N7 and N1. The overall yield for the whole production was about 50% of that predicted by theory.

The pure, dry product was then taken to the milling house, N12, where the picrite was reduced to a uniform particle size, a process which was very important in the production of a propellant with uniform ballistic properties. Stocks of picrite were kept in a magazine either at M12 or M13 in the northern quarter of the factory until required by the cordite section. The picrite section worked continuously twenty-four hours a day, throughout the Second World War in order to help meet the demands for flashless cordite.

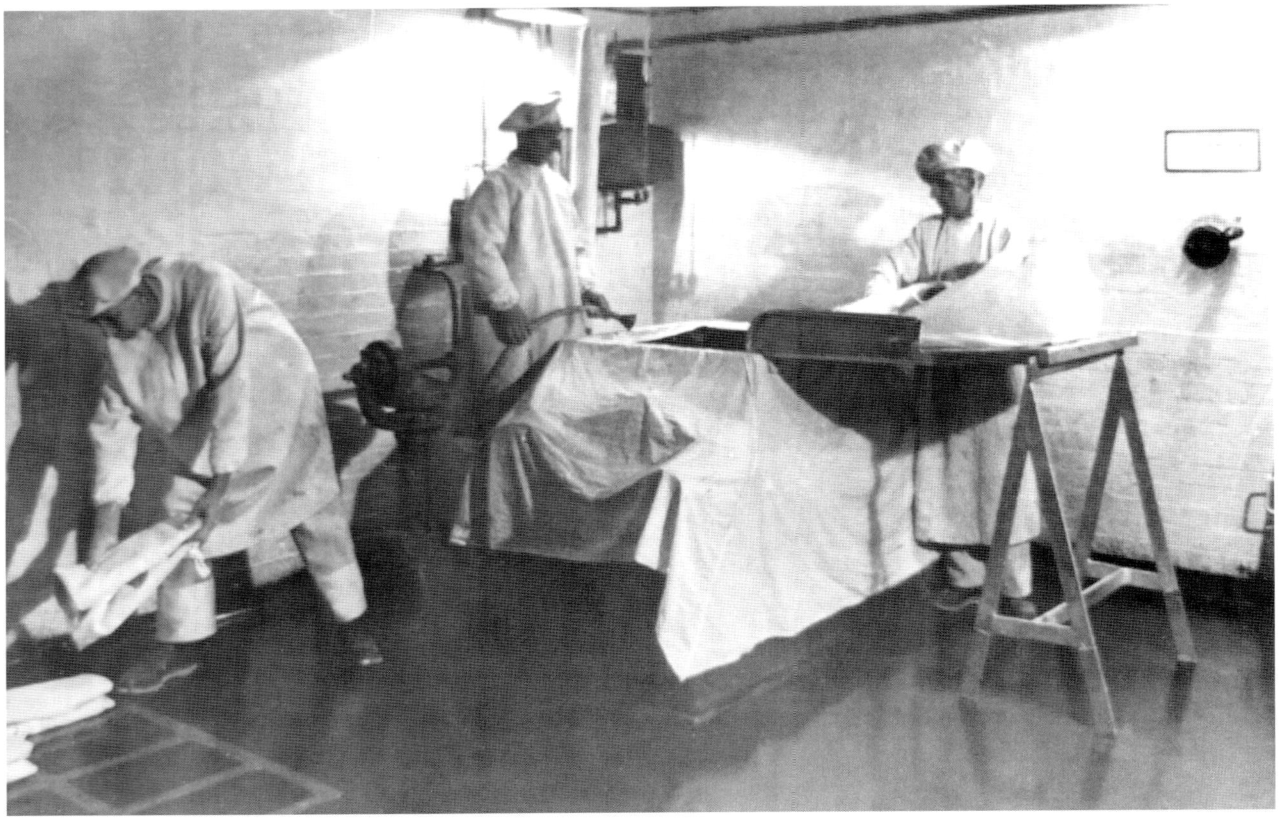

Left: This photograph shows workers spraying acetone on to a mixture of dry cordite paste and picrite, prior to incorporation. The vessel on the wall in the corner of the building is a burette to supply a measured amount of acetone.

Right: A progress photograph of N13 part of the picrite factory, under construction.

Left: An overhead view of the picrite nitrating vessels.

Tetryl

The tetryl plant was built at RNCF during 1916 and this high explosive was the only material to be deliberately produced at the factory which was not directly or indirectly connected with the manufacture of cordite.

Tetryl is a pale yellow crystalline material which was first prepared by Mertens in 1877. Its use as an explosive dates from the early years of the 20th century and it has been widely used since the time of the Great War as an intermediary or 'booster' (sometimes known as a 'gaine') between the detonator and the main explosive charge in shells and bombs. Some high-explosives for example trinitrotoluene (TNT) are surprisingly insensitive and require a bigger initiation than would be provided by a detonator alone. The intermediate tetryl gaine provided the 'amplification' of the detonator explosion.

Tetryl is known by chemists as 2,4,6-trinitrophenylmethylnitramine but in military parlance it becomes Composition Exploding, or even more succinctly, CE. At Holton Heath it was commonly known as tetryl and this is the name that is used throughout this account.

A batch-wise production method was used at RNCF, although a continuous process for manufacture was developed in the 1930s. The raw materials required included sulphuric and nitric acids, both of which were manufactured on site, acetone and dimethyl aniline (DMA). Acetone was made at the factory until the late 1920s, when the cordite manufacturing process changed but after this time the acetone for tetryl manufacture had to be bought in. During the late 1930s it was supplied in drums containing 700lb by British Industrial Solvents Ltd. The DMA, which is a dark coloured oily liquid of somewhat unpleasant odour, was also bought in and by 1937 the only supplier was the Clayton Aniline Company. The only other material used was distilled water, which together with acetone was employed in the purification process.

Stage one in the manufacture was described as that of 'sulphonation', although a chemist from outside the explosives world may object to the use of this word since the process was in fact that

Below: The tetryl nitrating house T1 in the foreground, and at the rear of the picture, the acetone recovery and purification buildings. Note the complex fume extraction towers with acid traps at the front.

of forming the soluble sulphate of DMA. Explosives chemists are, of course, well aware of this but the use of this word to describe the process is traditional and it has been absorbed into the jargon.

The sulphonator itself was a lead vessel fitted with cooling coils fed with cold water and with perforated coils supplied with compressed air to provide agitation of the reactants. A charge of 1,728lb of 94% sulphuric acid was first run into the sulphonator from the sulphuric acid store tank which was mounted on brick piers just outside the nitration house. This steel tank, which was 118 feet long by 5 feet in diameter, held about 50,000lb of acid which was sufficient for about thirty sulphonations. The tank was supplied to RNCF in 1916 by Messrs Piggott of Birmingham for the sum of £49 15s.

Compressed air agitation was then started and the 108lb charge of DMA was introduced gradually into the sulphonator through a lead injector. The temperature was kept below 20°C throughout. When the sulphonation of DMA was complete – and this took about two hours – the material was ready for nitration.

It will have been noticed that the process of nitration has been an essential stage in the production of all the explosive materials produced at the RNCF and it is also of fundamental importance in the production of many others as well, including such well known materials as TNT and Ammonite. The reason for this is that the so-called nitro groups introduced by this process are particularly effective in conferring upon the final product the qualities which make a material useful as an explosive.

John Read, in a most entertaining book on explosives, put it well when he wrote that:

'It is a curious fact that explosives are usually dependent for their existence upon the element nitrogen – an inert gas which enters into chemical combination only with great reluctance, and retires from such distasteful gregariousness into 'splendid isolation' at the first opportunity. The function of the nitrogen atoms is to hold away the oxygen atoms (or most of them) from the carbon and hydrogen atoms until the signal for molecular Armageddon arrives.'

Left: An 'F' house or drying stove. Originally designed for paste drying, this stove seems to have been adapted for picrite, since some of the trucks are labelled 'RDN'.

Right: An internal view of the tetryl factory showing a sulphonator with, in the centre, an early-pattern, mechanically agitated enamelled iron nitrator.

Left: This view shows a lead sulphonator situated in the tetryl factory, together with an adjacent nitrator unit.

The nitrator was made of lead and was similarly fitted with a means of air agitation and with lead coils through which either hot or cold water could be passed. The standard quantity of 8641b of 90% nitric acid was then run into the nitration vessel and warmed to 50°C. The sulphonated DMA was then injected with compressed air into the agitated nitric acid. As the nitration proceeded, so the temperature rose and by careful manipulation of the temperature of the water in the cooling coils, this was controlled so that it did not exceed 70°C. When the reaction was complete – and this took about two hours – the mixture was cooled to 25°C and the crude tetryl settled out in the form of yellow crystals.

This process proved to be a very reliable and safe method of nitration and while this equipment was used, there were no instances of ignitions. In earlier days, when enamelled iron nitrators were used, equipped with surrounding water and steam jackets and mechanical agitation, there were occasional problems with charges igniting but, fortunately, there were never any fatalities or serious accidents associated with them.

The next stage was that of separating the crude tetryl from the spent acid and this was achieved by filtration. The product was then washed with cold water to remove acid residues and then transferred to a vat in which it was boiled. This process served to remove various water soluble impurities which were generated as a result of side reactions occurring during nitration.

The boiling vat was of pitch pine with a spruce bottom. Heating was provided by means of coils carrying live steam, and boiling was continued for two hours after which the process was repeated with fresh water. The water was then drained away and the tetryl transferred to the basket of a centrifuge known as a wringer, which facilitated the removal of the bulk of the remaining water. After wringing the tetryl contained 5-6% of moisture and was ready for the acetone purification stage.

2751b of the partly-purified tetryl was then placed in an enamelled iron dissolving vessel equipped with a stirring shaft fitted with paddles made of beech wood and containing 6801b of acetone. The mixture was stirred until solution was complete after which it was passed through a filter and into a precipitation vessel, which was similar in structure to the dissolving vessels.

When all the solution had been filtered, 490lb of distilled water was fed into the acetone solution through aluminium jets at such a rate that the addition was completed in 230 minutes. The mixture was stirred throughout. No precipitation of tetryl occurred in the first 30 minutes or so of the addition but thereafter the pale yellow crystalline material continued to precipitate out until all the water had been added. The rate of water addition was critical because it controlled, to some extent, the size of the crystals formed. This in turn was important because crystal size had to be about right if the later sieving process was to be straightforward, and in addition the heat stability of the product was also influenced by this same factor.

Wet tetryl was then packed in bags and wrung, two or three at a time, in another centrifugal wringer. The tetryl was then transferred

Below: The tetryl filters. This photograph clearly shows the corrosive nature of the product at this stage, by virtue of the degradation of the ceiling adjacent to the fume outlets.

Right: The interior of tetryl nitrating house T1. This photograph shows the wooden-clad boiling vat with an aluminium lid and chute leading to the tetryl 'wringer', a belt-driven centrifuge. Also shown are the scales for weighing the crude product and a table for storing the tetryl.

Left: Tetryl crystallising vessels and filters.

from the bags with an aluminium scoop to American white wood drying trays, very similar in design to those used for cordite and lined with filter paper. The trays held 251b each.

The drying stove T3 was used exclusively in the early days of tetryl manufacture at Holton Heath.

Loaded trays were placed in racks within compartments in the stove. Air was heated to 70°C as it passed over steam pipes near floor level and was induced by means of an air jet above the trays to pass over the surface of the tetryl in the trays. The temperature was monitored throughout the 48-hour drying cycle at two-hour intervals, as the temperature was never to go above 80°C. The trucks used to convey the tetryl from one house to another travelled on 6 inch by 3 inch, oak rails when inside the building. After drying, the product was trucked to the sieving house, T4, and it was here that workers were exposed to the most serious health hazards in the whole manufacturing process.

The material was sieved to remove oversize particles and a fine dust of tetryl pervaded the atmosphere. This settled on the exposed skin of workers and frequently gave rise to a nasty form of dermatitis. To help reduce the incidence of these occurrences, a whole range of precautions were taken. All operators had to wear rubber gloves and a muslin mask which covered the lower half of the face. Immediately after a period of operation in the sieving house, all workers had to take a bath and a register was kept to show that this regulation was adhered to.

To wash ones hands after visiting the toilet is, of course, a very sensible hygienic measure for anyone to take but for the tetryl workers it was even more important to wash their hands before any such visit and there was a regulation to this effect. They were also required to wash their hands in hot water before eating and a special soap supplied by John Knight & Co was used for this purpose. Each tetryl worker was also supplied with two complete changes of clothing. Thus it can be seen that much was done to reduce health hazards to a minimum.

Individuals vary enormously in their susceptibility to the physiological effects of tetryl dust on the skin. In general, dark skinned workers proved to be less prone to dermatitis than those with fair skins. In any event, those working in the tetryl factory were given monthly examinations and those at any time showing the typical symptoms of skin blotchiness and swelling were instantly removed from the work. In spite of these precautions, no fewer than nine workers had to be permanently removed from work in the tetryl factory during the two years ending November 1939.

It was common for tetryl workers to have their skins stained yellow but provided that there were no indications of dermatitis, this was accepted as an unavoidable consequence of the work. Because nearly all surfaces in the sieving house became covered with a fine layer of tetryl dust, there was also an ever-present danger of a more immediate kind and the greatest possible care was therefore exerted at all times to avoid anything in the nature of mechanical shock or friction within the building.

A batch of tetryl was sieved in portions of 200lb and, after each portion had been sieved, it was packed in boxes and trucked to a short siding, T5, where it was temporarily stored until the whole batch had been sieved. When the sieving and packing operation was complete, the whole lot was removed to a magazine from whence it was ultimately dispatched to a filling factory. Some tetryl 'gaines' were clearly produced as well, since the factory included a tetryl press. The press die contained a central spigot which left the pressing with a hollow centre, into which the detonator would be placed. These were destined for depth-charges.

The output of tetryl from the factory varied over the years as might be expected. Production was obviously high during the First World War but thereafter output slumped. In 1937, 1,5001b per week was being produced with a staffing of a chemist in charge, a chemist, a foreman, an assistant foreman and ten workers. Six workers were employed on the day shift which operated from about 8:25am to 5:10pm and four workers on the so-called first shift which worked from 6:25am until 3:00pm. By 1939 output had risen to 4,000lb a week but by then 25 workers were employed on a three shift basis. To meet the increased needs for drying facilities, a new stove, T6, was constructed. By 1940, tetryl was being produced at the rate of 7,0001b per week.

Wages for tetryl workers were high by comparison with those earned elsewhere within the factory. In 1937 they earned £2 9s per week, plus a £2 bonus, for a 47 hour week as compared with £1 16s plus £1 bonus for guncotton workers and £2 2s plus £1 bonus for acid workers on similar hours. At the beginning of the Second World War wages were the same but the bonus figure had been raised to £2 2s in all cases. In addition, those working at night also received a night shift bonus. By 1940, wages for acid workers had risen to £2 2s plus £1 7s bonus for a 47 hour week and it may be assumed that the bonus paid to other workers would also have increased by 5 shillings per week.

Above left: Tetryl crystal centrifuge.

Left: A mechanical tetryl sieve. A mechanical tetryl sieve.

Above: An exterior view of tetryl stove T3. The sign above the door reads 'Limits: Explosives 1100lbs, Workers 2'. Note the light fittings located outside the building but shining in through the windows.

Above right: An exterior view of tetryl stove T3. The sign above the door reads 'Limits: Explosives 1100lbs, Workers 2'. Note the light fittings located outside the building but shining in through the windows.

Right: The final product! Cordite being weighed and boxed, ready for despatch.

Chapter 14

THE FINAL YEARS

After the cessation of manufacture at RNCF in 1946, Naval propellant requirements were met by the Royal Naval Propellant Factory (RNPF) at Caerwent. The output was very small and all gun propellant used was of the flashless type and contained picrite. No facility for making picrite existed at Caerwent and materials supplied from the Welland Works of the North American Cyanamid Company of Canada were used. In 1949 the requirement for propellant for the year 1949 to 1950 was doubled. In order to meet further increases in demand, two-shift working on gun propellant started in August 1951 and three-shift working started on 9th June 1952.

In view of the demand for picrite and the lack of a manufacturing unit at Caerwent, it was decided to re-open the plant at Holton Heath. In November 1949 Dr Charles Perrot, who had served on the nitroglycerine and acids sections at RNCF in 1939 and 1940 before being sent to Caerwent in late 1940, was sent to Holton Heath to start the project.

During the three years from 1946 to 1949 nothing had been produced at Holton Heath. The picrite plant had been on a care and maintenance basis during this time and manufacture started again in January 1950 using the plant without modification. The fusion of dicyandiamide and ammonium nitrate was carried out on a batch basis but soon a continuous fusion process was developed and put into production. This was a superior production method and became known, understandably, as the 'confusion' plant.

Bert Fricker – of whom we have already heard – was sent as foreman to assist Dr Perrot. Production was rapidly built up until a three-shift system was operating. All the picrite produced was sent to Caerwent. It is worth noting that around this time, a small ordnance factory, ROF Creekmoor, although having no connection with the RNCF, was closed. This factory manufactured small arms and was set up in the mid-1930s. It was located in the area where the Plessey factory more recently stood, off Soper's Lane.

The sulphuric acid required for the dehydration stage of the picrite production process was originally sent from Caerwent in railway tank wagons. RNPF wagons Nos. 17, 18, 19 and 02 worked a regular shuttle service. (The wagon really was numbered 02 and it is concluded that perhaps the person with the paint brush had made an error with the stencil.)

Manufacture of sulphuric acid ceased at Caerwent in 1952 and the Tentelev plant at Holton Heath was re-opened. John Gray remembers seeing Arthur Hunt of the laboratory staff weighing the platinised asbestos and packing it into the holders prior to placing these in the converters. The nitroglycerine mixed acid storage tanks were reconditioned for sulphuric acid storage.

The tetryl laboratory was re-opened to provide laboratory facilities for the picrite and sulphuric acid plants. To staff this laboratory, Arthur Hunt, 'Teddy' Terrell and Wilfred 'Pedlar' Palmer were sent from Caerwent, although all had previously served at Holton Heath. Other staff were recruited locally.

In 1952, to augment supplies of rocket propellant (SUK), a decision was taken to resume manufacture at Holton Heath. For this purpose paste sheet was made at Caerwent, dried and partly rolled after which it was conveyed to Holton Heath by rail. Rolling mills and horizontal presses had been rehabilitated so that rolling of the material could be completed and pressing carried out. The former nitroglycerine laboratory, Building X3, was refurbished for cordite work.

In 1955 a further development took place. A further need for propellant of the MNF type arose and a decision was taken to start manufacture at Holton Heath to augment supplies from Caerwent. For this purpose, wet paste was conveyed in a special container mounted on a road vehicle from the propellant factory to Holton Heath. The material was dried, incorporated, pre-pressed, rolled, pressed and dried at RNCF. All this required the rehabilitation of the appropriate buildings.

It was during this post-1949 period that it was decided to build a picrite plant at Caerwent. The construction of the plant, which was much bigger than the Holton Heath installation, went on from 1954 to 1955 and it was commissioned in 1956. The manufacture of picrite then ceased at Holton Heath, as did the manufacture of sulphuric acid.

The production of propellant was stopped in the summer of 1957 and most of the staff was dispersed, although a small number were retained to supervise demolition and removal of plant. During the closure period quite a lot of the plant was sold. A Mr W T Gray acted as an agent and was able to find purchasers. For example, the Tentelev plant was sold to a Canadian firm and sent to Liverpool for shipment in sections. Spare nitric acid stills were sent to India where nitre distillation was still practised. Sadly, Mr Gray later lost his life in an air accident. The Bowman Engineering Company carried out a lot of the removal work involved in the final stages of closure.

Dr Lee Smith, who had been Superintendent since 1952, retired on his 60th birthday on 9th October 1959 and thus ended the line of chemists who had been responsible for the production of explosives at the RNCF since it opened a little more than 43 years earlier in the middle of the Great War.

The very last accident which was recorded at the cordite factory happened when Mr Edward Charles Amey, a skilled labourer, was struck by a railway sleeper falling on his left foot. The date was 17th June, 1959.

Meanwhile, a significant development had occurred in 1947 when the Admiralty Materials Laboratory was formed and staff occupied redundant buildings within the factory area on either side of the road leading to the station from the A351. The formation of this research and development organisation enabled some staff employed at the factory to transfer over and thus continue to be employed at the site. This laboratory existed at the site until 1991 when all operations were gradually moved to other locations. During this period it bore the subsequent titles of Admiralty Materials Laboratory (AML), Admiralty Marine Technology Establishment (AMTE), Admiralty Research Establishment (ARE), Defence Research Agency (DRA) and lastly, Defence Evaluation and Research Agency (DERA). The site remained in MoD hands on a care and maintenance basis for some years after closure, with a commercial security firm in charge until eventually being sold.

As a Research and Devlopment establishment, the Holton Heath site was dedicated for many years to providing a service exclusively to the Royal Navy, and this continued until the amalgamation of the UK non-nuclear R&D organisations after which, with the formation of the DRA, it provided tri-service support.

Whilst security remained paramount in relation to defence materials research, open publication of security-cleared work was encouraged. Thus, the atmosphere was more relaxed and in sharp contrast to that at the RNCF during to the frenetic war periods when staff who worked at Holton Heath during the Second World War were forbidden to even mention where they worked, let alone the nature of their work.

Even at the RNCF however, publication of security-cleared work was relatively common during the inter-war period and after the Second World War. Microbiologists, Thaysen and Bunker, who were closely associated with acetone production at Holton Heath, were publishing in the open literature during the early twenties for example. Similarly, Alan Parker, a distinguished chemist at Holton Heath, was publishing work of general interest, but related to his more specific RNCF research, in 1948. Latterly, DERA sought commercial contracts in order to minimise the burden on the public purse and in those circumstances open publication of research was understandably influenced by the commercial interests of the customers, though scientists were still encouraged to publish defence-related papers of an unclassified nature.

When the site closed most of the DERA staff at Holton Heath were transferred to either Farnborough or the more local Winfrith site (formerly the Atomic Energy Establishment). In 2001, DERA came to an end under the Blair administration, and was split into two parts, the first being a privatised company (QinetiQ) whilst the second became the Defence Science and Technology Laboratory (DSTL) and remained within the Ministry of Defence.

Fortunately the site, despite its chequered history, has been well maintained, in a marked contrast to the unused land adjoining, some of which has re-grown to a such an extent that John Gray, on a return visit in 1987, had to refer constantly to the site plan despite having worked for so long at the factory.

At the cessation of cordite manufacture smaller parts of the factory were leased to commercial organisations including the Overhill Engineering Company (September 1956) and the Decca Record Company (July 1961). Both of these firms have now moved on but, for a while, some former RNCF staff were employed by Decca whose business was, of course, pressing vinyl rather than cordite. The huge main boiler house remained in use at this time and many rail travellers will recall seeing the large poster near the boiler house proclaiming 'Decca ffrr – full frequency range recording.' One of us can well remember a happy occasion when, as a part-time musician – and a bass guitar player for 'Bill Elliot & his Band' – he was contracted to play at the Decca canteen which was probably at one time the acids mess and changing room.

As that part of the site designated as an industrial estate gradually became populated, many of the old cordite buildings were demolished to make way for modern factory units. One set of buildings which survived with a new use for a considerable time was the group comprising the pulping and moulding house and the main boiler house. These were occupied for a considerable period by a firm called Fibaflow. Observers may well have reached the conclusion that the output of this company consisted of huge drain-pipes, given the number of these stacked in the yard between the buildings. Enquiries, followed by an invitation to tour the works, showed that in fact these were spun nylon tubes and their destinations were to be the cores of MRI body scanners – not drains! Sadly a recent visit showed that these buildings have now been replaced with modern ones.

It was inevitable that the end of war-time requirements would lead to closure of the Royal Ordnance Factories and to the demise of large-scale production at Holton Heath. Towards the end of the manufacturing period it was a common joke that the factory would spend a month producing cordite followed by a period of burning the old stock. There is of course some truth in this, since as with any such product, cordite has a limited storage life. This anomaly was given some publicity by a local resident, Mr Howard Gee. Howard was well known as being a thorn in the flesh of authority – an attitude which probably derived in part from his incarceration in Colditz Castle during the Second World War. Howard lived out on the Wareham heath in an old building known asDecoy Cottage. One of us has had sight of a number of old letters from Mr Gee to those in authority. The language used was colourful to say the least, even by the standards of today.

On the 18th December 1962 an agreement was drawn up for the sale of the remainder of the RNCF site, amounting to about 467 acres, back to the Lees family of Post Green, Lytchett Minster, the original owners of the land, in the name of Sir Thomas Edward Lees. The land subsequently passed into the hands of City Wall Properties, a development company taken over in 1971 by the Rank Organisation. The owning company is presently British Land. Since then a proportion of it has been committed to industrial development but a larger proportion, amounting to 290 acres, has been made available to the Nature Conservancy Council, later re-named as English Nature and more recently Natural England. The organisation now has a lease on this land to run for 999 years at a peppercorn rent.

Left: A shift clocks off from the RNCF. The building to the left of the gate housed the hospital, while that on the right housed the site police, a use which was maintained until the MoD vacated the site.

Chapter 15

THE MANUFACTURE OF EXPLOSIVES TODAY

Although our story must finish as far as explosives manufacture at Holton Heath is concerned, it is interesting to follow the path of the government's explosives manufacturers after Holton Heath closed and to take a quick look at methods in use today. One remarkable fact is that cordite has not been replaced as a propellant for a great number of uses and the basic chemistry of manufacture has not altered, although manufacturing methods and safety have improved. Cordite is often now produced in a cast, solid form specifically for rockets and missiles in addition to the 'spaghetti' type bagged cords which are still in use for conventional Naval gunnery.

During the period 1930-40, the 'state of the art' explosives manufacture which had been built up at Holton Heath was made available to those engaged in propellant manufacture in this country and in the Commonwealth. In 1936 it was decided that a further factory would be required in addition to Holton Heath. Thus the Chief Superintendent (Armament Supply), H J Burnett-Hall, requested the Superintendent of the RNCF to define the site requirements for a new cordite factory capable of a maximum output of 150 tons of cordite per week.

After various sites had been investigated, the building of the Royal Naval Propellant Factory (RNPF) at Caerwent in Wales was commenced following an initial meeting on 31st January 1938, with the consultant engineers Freeman Fox & Partners, to acquaint them with the features required for the new factory. The RNPF, therefore, became the true 'child' of the RNCF, since it was designed by the staff of RNCF and involved the same consultant engineers. In addition, the RNCF supplied many key staff, for example Dr R S G Knight, who became Superintendent of RNPF in 1939. During its life, the factory had a unique safety record and no deaths by explosives were ever recorded. It is clear that all the valuable experience gained at Holton Heath was put to very good use. The RNPF was initially envisaged as a 500-acre factory, however, when potential expansion

Below: The buildings visible in this photographs are believed to principally comprise the Caerwent acids factory.

was taken into consideration and a later decision was taken to build two identical factories in order to safeguard against aerial attack, the land eventually acquired was closer to 2,000 acres.

A point worthy of note is the consideration shown to the local population in regard to access to established ancient buildings and beauty spots in the affected area. The official record shows that after strong representations by local people, a popular local schoolmaster was allowed to retain the small part of the site upon which he dwelt. We are rightly proud of the advances that have been made in regard to planning and local representation in later years, but one is tempted to wonder if, for example, the same consideration will be shown to those unfortunate enough to live near any modern large project.

The RNPF differed from the RNCF not only in size but also in the use of more modern plant and the fact that the principal site transport was by road instead of narrow-gauge railway. At Caerwent, the only railway in use was a standard-gauge connection for bringing in raw materials and dispatching the finished product. A system identical to that employed at Holton Heath for manual movement of hazardous material was the use of rubber-wheeled trolleys in concrete grooves. At one point on the Caerwent site can be found a veritable 'Clapham Junction' of sidings and points, to enable trolleys to pass or re-route. The site also differed from Holton Heath in that the transport was mostly at normal ground level, whereas at Holton Heath most of the rail tracks ran in cuttings. The RNPF was capable of manufacture of all the materials produced at RNCF and the regulation and operation of the site followed the same efficient naval pattern.

Despite its comparative youth, a decision was taken in March 1965 to close the RNPF and to concentrate all production at the Royal Ordnance Factory at Bishopton, Scotland, and this finally occurred in 1967, thus ending the period during which the Royal Navy had exclusive control over its propellant manufacture and had built up a reputation second-to-none for quality and safety standards.

The site at Caerwent was for some time run by the Royal Air Force as a store in conjunction with the United States Army. Most of the original buildings and earth traverses are retained, although all

Below: A view thought to be of the nitrocellulose production group.

plant has been removed. One of us had the pleasure of visiting the site in October 1987 in the company of John Gray and the Commanding Officer, Squadron Leader Dave Chadwick.

At the time of the 1987 visit the first things one noticed about the main office were the posters referring to protection of wild life. It seems that Caerwent, like so many military sites (including part of Holton Heath today) is a refuge for endangered species of flora and fauna. We are indeed fortunate that the military take such an enlightened attitude to conservation and still reserve large tracts of land. Another good example of nature conservancy as a consequence of military occupation and restricted access, are the lands reserved by the Army on the Bovington and Lulworth ranges and at Tyneham village.

It was most interesting to see the design of the Caerwent site and how it had developed from the previous experience at Holton Heath. The factory appeared very open and exposed compared to Holton Heath, mainly due to the surface roads instead of deep railway cuttings for transport. In addition, the layout is very clear and uncluttered, as the site did not gradually develop in the same way as did Holton Heath. This must have been something of a camouflage problem during the war, although the RNPF was never seriously attacked.

Around August 1993, Caerwent ceased to be used as an active MoD site, although some minor activity took place during the Iraq and Afghanistan actions. A second visit around 1995, which was kindly arranged by Medwyn Parry of the RCHMW, showed the site to be somewhat overgrown and with one or two structures destroyed as a result of training exercises. The site at that time was used sporadically for Army and Police training (as was Holton Heath). In addition some parts were used for commercial storage and various open parts were grazed. The site owned by the MoD has currently been reduced, with some land having been sold, but a large part of the site is retained for training operations.

It was felt that this account could not be completed without a visit to a modern plant in order to note how practice differed from the Holton Heath methods. After some discussion with John Gray, it was decided that an enquiry be raised with the staff at ROF Bishopton. A visit was subsequently arranged through the kind offices of the Assistant Director, Mr Jim Wallace, and Les Hayward was shown around the factory by Mr Wallace and Mr Hamish

Left: The junction in the distance is where the railway branch that serves the factory connects with the nation network.

Left: This photograph shows the locomotive shed at Caerwent with an abandoned MoD diesel locomotive still in residence.

Below: This building was probably a rail unloading facility. Note the words engraved over the top of the middle opening: LOCOMOTIVES NOT TO ENTER THIS BUILDING

Christie – a chemist of considerable experience and repute in the field of nitroglycerine manufacture. Sadly, Hamish died not long after the visit.

It is worth noting that John Gray, who has contributed much to this account, worked not only at Holton Heath and Caerwent but also at the Royal Ordnance Factory from whence he retired as Assistant Director (Research), having been responsible for the development of combustible cartridge cases which disintegrate in the firing chamber, thus resulting in much saving of material, weight and clutter.

The visit took place on 16th July 1987 and the first impression caused two surprises. Firstly, the size of the factory site, which is about three times that of Holton Heath and secondly, the presence of the old plant which remained along with the new and which, although now unused, provided an interesting link with the past. It was somewhat strange to see equipment and buildings still standing which had previously been seen only as in early progress photographs from Holton Heath. There were still, for example, buildings which once housed Nathan-Thomson-Rintoul batch process nitroglycerine plants and one building which contained waste acid 'eggs' of a similar type to that involved in the 1931 explosion at Holton Heath. Another example of the past coming to life was the sudden view of old rolling houses identical in design to those which once stood at Holton Heath.

Although the ROF once used Schmid nitrators identical to those

Below: This building once housed the nitrocellulose slurry tanks.

Above: A view inside the deserted nitration house for nitrocellulose at RNPF Caerwent. Note the fine ceiling and also the remnant of Police tape, indicative of the building's later use for training purposes.

used in AB2 at Holton Heath these had been replaced, but in other respects and in many other aspects of explosives manufacture, the ROF remained a treasure house of industrial history. The reasons for this are not altogether altruistic, as in some cases old plant and buildings had been retained for other uses and others had simply remained because that was easier than decontamination and removal. There was, after all, no shortage of room and the site was sufficiently secured.

The next generation of nitroglycerine plant at ROF was to be the Swiss Biazzi continuous plant Swiss, of which two were employed at the ROF. This very elegant apparatus well illustrates two main trends in the manufacture of nitroglycerine, namely that of reducing the complexity of washing and acid recovery and most important for safety, a reduction in the amount of nitroglycerine accumulating at any part of the manufacturing cycle.

It will be recalled that the original batch process used a nitrator-separator and pre-washer within the AB house, followed by a complex washing process in a completely separate plant, the 'C' house. The washing stages with the Schmid process were also very involved, using several washing columns and intermediate separators in both the nitrating and the washing houses. The original quantity of nitroglycerine manufacture using the batch process was over a ton. The resulting explosive force of such a quantity of nitroglycerine was well illustrated by the 1931 Holton Heath disaster and, apart from the convenience of operating continuous plants such as Schmid and Biazzi type, safety is clearly improved by having far less explosive in one concentration. A further improvement in safety is the transport of the finished nitroglycerine which is now piped in a non-explosive mixture with water directly to where it is required, namely the mixing house. The only time nitroglycerine is carried about by manual means is when a small quantity is removed for analysis.

It is interesting to compare first the Nathan-Thomson-Rintoul

Below: The main boiler house and sub-station at Caerwent.

Above: The purpose-built magazines at Caerwent, designed to house American ammunition moved here from temporary storage at RAF Fauld following the removal of NATO stores from France.

batch process with its vast nitrators and separators with the Schmid plant, smaller and capable of continuous output but requiring fairly extensive washing plant and then the Biazzi, which is a long and very narrow plant, occupying one wall only. Materials used in the plumbing have also changed from lead to almost exclusively stainless steel, making the Biazzi plant a very smart piece of apparatus. Plastics are used to advantage for corrosive liquids.

Much use of the modem art of process control had been applied to all plant at the ROF and the Biazzi plants were controlled from a separate control room while nitrating, giving operators complete safety. It is worth noting, however, that any of the plants mentioned here were capable of perfectly reliable and safe production if used correctly and with good, clean material. Mistakes and unforeseen events will always occur, however, and it is far better that a small amount of nitroglycerine explode harmlessly in a remotely-controlled plant, than if one were to be blown up working alongside. John Gray was not entirely convinced that personal inspection of operating plant is unnecessary, however, and cited an incident in which a personal inspection revealed a leaking pipeline spraying a wooden beam with acid, an event which was not visible on the television camera and which could easily have given rise to fire. On another occasion, an actual fire was found to be in progress, after a rubber seal on a light fitting had come into contact with the hot bulb.

The latest plant to be employed at the ROF was of the NAB 'Gyttorp' type, named after a place in Sweden. This plant is also of the continuous type but instead of passive separation it uses a three-stage centrifuge process. The plant may be operated fully automatically by computer. The ROF plant was noted for its simplicity of construction, making it appear a rather boring piece of apparatus compared to the photogenic Biazzi.

The plant was housed in small adjoining concrete houses each containing the nitrator and the centrifuges in isolation. This limits the damage to one small area in the event of major failure, as the individual buildings are of sufficient strength to contain the worst explosion that might occur. In addition, the maximum explosive material is reduced by a factor of ten by comparison with the Biazzi, and most of the nitroglycerine in process is in the form of an emulsion which cannot detonate. A major feature of the plant is the speed with which it can be started and, more importantly, stopped. A number of serious accidents have occurred in the past with Schmid units which could have been prevented had it been possible to stop the reaction quickly.

The transport of nitrocellulose to the mixing houses at the ROF was also accomplished by means of water, in a similar manner to the last method, the ring main system, pioneered and employed at Holton Heath. Nitrocellulose can, of course, be as dangerous as nitroglycerine when dry. The final stages of cordite manufacture were also improved at the ROF where they had developed a method of incorporating, rolling and pressing not only in one house, but in one machine. Hitherto, this process would previously have required three separate buildings, plus transport in between.

Sadly, even the efficient Bishopton site eventually closed as an explosives and propellant manufacturing site. The expensive new NAB nitroglycerine production unit shut down around 1990 after a very short life and only a small part of the site is now in use at the time of writing by the owners, BAE Systems, as a test and evaluation facility. This demise follows the same sad story of British explosives manufacturing which includes the closure of ROF Bridgewater in Somerset – a somewhat surprising development given that a brand new boiler house there had not long been in service and the factory was producing some popular items including 'Blade', a shaped plastic charge popular with demolition contractors. Enquiries made at the time of writing this second edition show that in fact the production of propellants is barely supported in the UK. The Chemring Energetics Company at Ardeer produces Mechanite, a solid Extrude Double Base Cordite

Left: Looking from behind the portal into the short tunnel which once formed the entrance to the Schmid nitrator house.

Below: A photograph showing the overgrown asphalt tramway which was used to guide the nitroglycerine trolleys from the nitrator.

Opposite: A view of the remains of one of the nitration units. It was originally built on the same principle as AB2 at Holton Heath, and would have contained Schmid nitrators. The building and its earth cover have been removed as part of the de-contamination measures.

for guided weapons, as does a company called Roxel. Of conventional cordite manufacture there would seem to be none in the UK, despite the fact that the material is still in demand for ships' gunnery.

As in all industry, as efficiency of manufacturing is improved, fewer people are required to do the same work and while this is clearly a desirable trend it is, perhaps, unfortunate that the ROFs will cease to be such a significant employer in areas which could well benefit from some extra jobs. Part of Holton Heath continues to be used as an industrial estate and although it is very unlikely that the original employment opportunities will exist again, a significant number of jobs are becoming available at this once deserted site, while on the eastern side of the site the remains of the old drying stoves will continue to provide a refuge for wild life under the watchful eye of the conservancy bodies.

This reserve was opened officially by Lord Cranborne, the MP for South Dorset on Wednesday 23rd September 1981. Under the oversight of the Nature Conservancy Council warden Mr Mike Tuck, large areas of dense undergrowth were cleared to assist the wild life and promote the re-growth of heather, which incidentally restored some of the grand views once enjoyed by the cordite workers.

Many people associate explosives exclusively with war and guns. While it is true that munitions probably do represent the most significant market for explosives, it should be borne in mind that explosives are also essential for use in mining and demolition operations as well as in a wide range of safety and potentially life-saving equipment such emergency flares, acoustic warning devices for railways, and, probably the most dramatic of all, the rocket motors used in aircraft ejection seats, pioneered by the British firm of Martin-Baker, whose equipment has saved many lives.

Below: This view of the nitrocellulose factory also demonstrates the wide separation of adjacent buildings. Although the hand-rails indicate that the corridor roof was intended as a walkway, it would probably have been for maintenance access, since all processes were contained within the buildings.

Above: A general view of the RNPF Caerwent. This picture was taken from the roof of the nitrocellulose factory. The figures in view are Nick McCamley and Medwyn Parry.

Above: An example of current nitroglycerine manufacture. This is the 'Gyttorp' plant at the former Royal Ordnance Factory Bishopton, Scotland. The system is designed such that due to the minimal amount of explosive material present at any one time, a possible explosion may be contained within the concrete enclosures. The large building contains the nitrator and the three to the left house centrifugal separators.

Chapter 16

THE ROYAL NAVAL CORDITE FACTORY ASSOCIATION

As it became clear to all that the MoD would eventually be withdrawing their interest in Holton Heath, DERA staff, the local authorities and of course the writers of this document, realised that a significant part of the history of Dorset was in peril of being lost. Preparation for publication of the first edition of this book in 1996 had also ignited considerable interest from the general public and former workers at the plant, and they, and in some cases their descendants, had come forward to either add snippets of information or to ask questions about aspects of the work which had puzzled them for years.

As a result of this interest a number of visits to the site were arranged with the blessing of DERA and with the help and organisational skills of one remarkable lady, Mrs Maureen Ashe. At the time of DERA occupation, Maureen was the head of site services and was thus in a position to be able to make the necessary contacts to facilitate such visits. A further boon was that at the time one of the authors, together with John England and Robert Honeybun, still worked on the site and so a number of supervised and low-key visits took place, normally with John England and Bob Honeybun as 'minders'. Les Hayward was not then employed by the MoD but was afforded the greatest possible assistance in visiting from time to time including a great deal of help from the MoD police – to the extent that on one occasion when escorting some members of the public he was made an honorary member of the police for around three hours, as all the regular staff were busy. It is an indication of the trust which had been built up that he was handed a radio and asked to supervise the visit himself! One of the site police was a former Royal Marine, a very helpful chap called Dave Mortimer who later joined the staff

Right: A view of part of the former museum at Holton Heath, which eventually occupied three rooms! This area includes the scale model and Second World War posters.

Right: A view of building D3 taken in 2003. As this was in use up until the site finally closed, it is a fine example of how an unattended building quickly reverts to its surroundings.

at the Winfrith site, he also later became a member of the RNCFA and assisted with traffic management at the open days.

Given the number of ex-staff either from the days of cordite or AML who expressed an interest, it was decided to arrange a small reunion on site. The venue for this was chosen as one of the old and now disused canteens. Preparation for this event required a very early arrival during the hours of darkness. Fortunately arrangements had been made to restore power to the building for this occasion. A good day was had by all, with old acquaintances being renewed and old memories and stories swapped.

Regular meetings had been held with the local authority (Purbeck District Council) and it was a fortunate coincidence that Les Hayward was, at the time, the Chairman of the Planning Committee and thus attended in this capacity as well. With the emergence of the North East Purbeck local plan and proposed housing increases which had been mandated upon the county, the District Council were taking a keen interest in the future use of the site and were just as keen as the individuals mentioned above that some work be undertaken to preserve the heritage and history of the site.

It became clear to all parties that not only were steps necessary to record and preserve aspects of the site's history, but all concerned were rather taken aback with the amount of interest and support which had been shown. As work from Holton Heath was gradually being transferred to other sites, space within the buildings at Holton Heath was being freed. At one meeting it was suggested that given the renewed interest it might be worth an attempt to recover the large scale site model which was currently housed in a leaking shed at Priddy's Hard at Gosport and then set this up in a spare room of the administration building. This was eventually achieved and proved to be a valuable talking point during a visit to the site by Ian Bruce, the Member of Parliament for South Dorset.

On arrival the model clearly displayed the signs of the years of indifferent storage and was much in need of some restoration. At the time, this model had also been recognised as a 'document of national importance' and, as such, a legal obligation existed for its care. As a matter of good fortune a voluntary conservation effort was put in by a team led by Anne Henderson whose father was formerly employed at the RNCF and who conveniently lived in one of the RNCF cottages nearby. Anne went on to offer continued assistance to the RNCFA.

The birth of the association took place at one of the regular liaison meetings which were taking place. It was suggested that such

Above: An example of the many buildings on site utilising the 'Belfast truss' type of roof. This picture from 2003 shows the signs of abandonment. Originally finished in plain brick, most of these buildings were later painted.

an association could not only protect and record the site history, it could furthermore bring more old workmates together, as had the successful reunion. Given that a small collection of artefacts was starting to appear on site, either from things that employees had kept 'in the drawer' for many years or which had come to light as buildings were being cleared, a museum was suggested with the newly acquired model as a centre piece.

A new home was found for the model on the first floor of the administration building and, largely through the efforts of John England, other rooms on the same floor were, by some means not explained, acquired, to store a growing number of artefacts. By this work an interesting museum was gradually created whereby the various rooms encompassed various examples of work undertaken during the whole life of the site.

Some discussion arose as to what the association should be called. A question was raised as to whether it was permitted to have the word 'Royal' in the title. Fortunately this question was quickly answered by Paul Croft, the Chief Executive of the District Council, who had attended various meetings and, being a solicitor, was able to give much helpful advice. Also present at most meetings was Dr David Evans, a senior planning officer. The help and advice these gentlemen provided proved most valuable.

The decision having been taken to create the association, it was noted that should the model be taken under the auspices of the association there would be a legal responsibility for care and preservation of it. It was Maureen Ashe who suggested that were we to take the burden of this from the shoulders of the MoD it should come with a 'dowry' in the form of some cash which would help kick-start the association. Further funds would come from members' donations and fund-raising events. Ever the optimist, Maureen wrote to the Chief Executive of DERA, Sir John Chisholm, who responded with a donation to the association of £1,000.

The association proved to be a great success, attracting to its membership, not only former employees but also others with an interest in local history. Among the non-employee members was Dr John Beavis who went on to become Chairman of the association. John's skills were particularly welcome, since his speciality was industrial archaeology in which he lectured at Bournemouth

Above: This building, which originated as the cordite office, was last used in the DERA years and housed the laboratories and office of Malcolm Bowditch. This photograph was taken in July 1997, after closure.

University. This connection also resulted in some excellent events at Bournemouth including a memorable visit by Dr Sidney Alford, who spiced up the talk with a few well staged explosions! Later on the Chairmanship was assumed by Dr Bob Dukes who was a former business manager of the DERA establishment.

When the DERA move from the site was complete, the establishment was put on a care and maintenance basis. The MoD police who had previously been responsible for site security took up new work at Winfrith. They were replaced by a private security company based on the ground floor of the administration building. During this period, right up to the time of site disposal, the association was very well supported by the MoD. The security force was given instructions to allow association committee members access at any time. The best and largest room was made available for the museum and this was redecorated by members before the exhibits were set up. Advantage was taken of this freedom at the time to introduce popular open days and guided tours around all parts of the site, accompanied by members of the committee. One memorable tour was arranged by Les Hayward, especially for some of the 'regulars' at the Fox Inn, Corfe Castle. The occasion was rather delayed for some reason into the late evening and darkness was quickly falling, with the result that a large part of the tour of the deserted buildings was undertaken by torchlight and in the rain. Thinking that this had been a bit of a disaster, it was amazing to hear later remarks to the effect that it had been a most enjoyable and interesting trip. Clearly, poking around dark buildings at night had also added a spooky atmosphere to the event!

During this period certain buildings on the site were used for

Above: The same building a few years later with Mother Nature having taken a strong hold.

police training exercises. Sadly, such training seems to always involve a certain amount of damage. A further unusual use of the site was as a setting for one of the *Bad Lads Army* television series. Although Holton Heath had no connection with the Army, some of the buildings presented a suitably military appearance to pass off as a Second World War Army camp.

Sadly, all good things must come to an end. This well-known phrase can also be applied to the RNCFA. A major factor in its decline was the fact that the MoD had sold the site. This was clearly the eventual outcome and the association had been very grateful for rent-free heated accommodation over a number of years. The site had been bought by a local developer, Birchmere Ltd. Fortunately the company was very keen on the history of the site and immediately carried out sympathetic restoration to a number of buildings, preferring re-use to demolition. The RNCFA were allowed to continue operations and some open days, but eventually it was clear that a move was on the cards since the administration building was the finest on site and the developers needed to employ this to get a return on their investment. A generous offer was made to re-house the museum in another building, however the services had long been removed from it and the cost of re-connection and the ongoing cost of heating and maintenance were beyond the modest means of the association.

As a result of this, and also the fact the association had been steadily losing members as a result of inevitable old age, the newsletter of summer 2008, edited, incidentally, by Malcolm Bowditch, carried an invitation to a last reunion in October of the same year, this time to be held at the Hamworthy Club. The event was to coincide with the Annual General Meeting and a proposal on the agenda was that the RNCFA should wind up its affairs. After some discussion it was seen that there was little other option and so the association was regretfully closed down and the museum's records put into the care of the County Archive service. The large factory model was returned to the *Explosion!* museum at Priddy's Hard, where by this time a proper public exhibition centre had been set up on site.

It should be noted that had the North East Purbeck local plan been implemented in full, the museum would have, in some form, remained. The plan called for housing development on part of the site with the area around the administration building being retained and the building itself passing from MoD to local authority ownership. The plan envisaged retention of the museum as part of the scheme, and that the surrounding area would become a form of community central park. Members of the RNCFA had been in close touch with the conservation group who took over the Waltham Abbey establishment just after closure. Waltham had received a very large grant for exactly what we were trying to achieve for Holton Heath, but there was little chance of a grant in our case. Waltham Abbey had certain clear advantages; one being a far longer history going back to the days of gunpowder, another was the fact that the site was intact. A further consideration was that it was close to the London conurbation. Had a museum been established at Holton Heath on a permanent basis, Waltham had indicated that they could make available a few items such as surplus cordite presses for outside exhibits. There were also a few items at Caerwent which might have been made available.

Former members of the RNCFA continue to produce documents, arrange talks, and have also produced an excellent professional video recording and website charting the history and manufacture of cordite, featuring members of the RNCFA as presenters.

The association, although having a relatively short life, achieved a number of important objectives:

- It brought together many ex-employees of the factory and its successors.
- It established an interest with the authorities, which led to statutory protective measures relating to buildings of special interest and to the collection of artefacts.
- It provided a great deal of pleasure for members of the public to enjoy the talks and open days and enabled them to ramble around an interesting and attractive site.
- It provided the means to make local people aware of the important history of the site.

Birchmere Ltd have done an excellent and sympathetic job with the re-use and letting of the site buildings. They have preserved the park-like atmosphere of the area and maintain it well. Sadly the former climatic laboratory area on the opposite side of Station Road is abandoned and derelict at the time of writing.

It would be inappropriate to leave this chapter without mention of the many members of the RNCFA and the public who assisted in some way with this project. Their names would be too numerous to mention here. Thanks must of course also go to the wives and family members who took on the onerous task of providing excellent catering for hundreds of people at a time during the various events and who assisted with car parking and generous stewardship.

It might be thought that a further record should be undertaken to detail the later work, after the closure of the cordite factory, that went on at Holton Heath – certainly there would be some good stories in relation to staff and social activity. However the later site work did not deal with a single manufacturing process of general interest, but research and testing procedures – some long term, some short. There may be a story yet to emerge from this period, but it would need to be diverse and possibly too technical to appeal to all. It should also be remembered that much of this work probably remains classified.

Below: The authors with two 'old hands' of Holton Heath celebrate the restoration of a fine cast iron lamppost made by Davey Paxman & Company of Colchester. This was recovered from building X3.
Left to right: John Gray, Malcolm Bowdith, Ted Cake, Les Hayward.

Chapter 17

SURVIVING STRUCTURES

Below: This photograph shows a group of visitors exploring the remains of the Second World War brick housings for the large mixed-acid tanks which supplied the Nitroglycerine Hill area.

Above left: The entrance to P16 press house around 1990. This became flooded due to earth works on the DRA site and a contribution from a leaking water main added to the problem.

Above centre: An interior view of a room in P16 - note the asbestos-lagged steam pipes still in situ.

Below left: The remaining tunnel entrance which led to nitroglycerine washing house C2, with electric switch gear housing at left.

Above right: The hydraulic pump and electric switch gear housing at the rear of P16.

Below right: The heater house and overgrown marshalling yard of picrite drying stove N7. Compare this with the picture on page 185, showing the plant in use.

Left: The Press House P17. Although very overgrown, P16, P17, and P18 were at the time of photography (around 1997) in a very good structural condition.

Above centre: The rubble which once comprised picrite house N13. This and other buildings in the picrite area were destroyed for no apparent reason. The picture on page 187 shows N13 in happier times.

Above right: The interior of the control trench. This should be compared with the view on page 68. Remnants of bomb-plotting charts were recovered from here by members of the RNCFA and reassembled.

Opposite below: This view shows the engineering shops to the left and the timber store to the right.

Right: A fine example of a magazine stores at the back of the laboratory.

Opposite above: A view of some of the remaining ancillary buildings looking towards the police gate from the station. These buildings in later years were used by the research agencies and housed various offices,reprographics and electron-microscope facilities. A wonderful environment for a stroll during summer, but walking from building to building in the rain of winter whilst clutching documents must have presented some problems.

Opposite below: This rather grand building houses the main laboratory.

Below: The photograph below is of the cordite measuring room.

Chapter 18

EPHEMERA

Right: An example of the special tickets issued by the contractors who built the RNCF. This was found with several others in one of the RNCF cottages when it was being cleared out. It is interesting to note that the maximum penalty for misuse of the ticket was £100, a very considerable sum of money at the time.
Below left: The front cover of the general and special rule book. All workers were given a copy and all were bound by the rules.
Below right: A rather long-winded poster offering a reward for detecting trespassers.

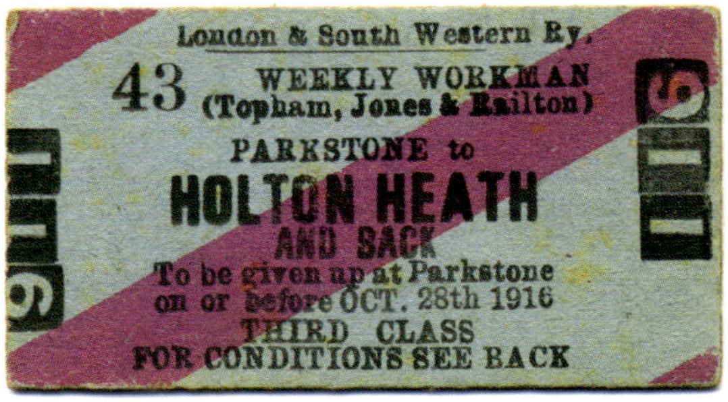

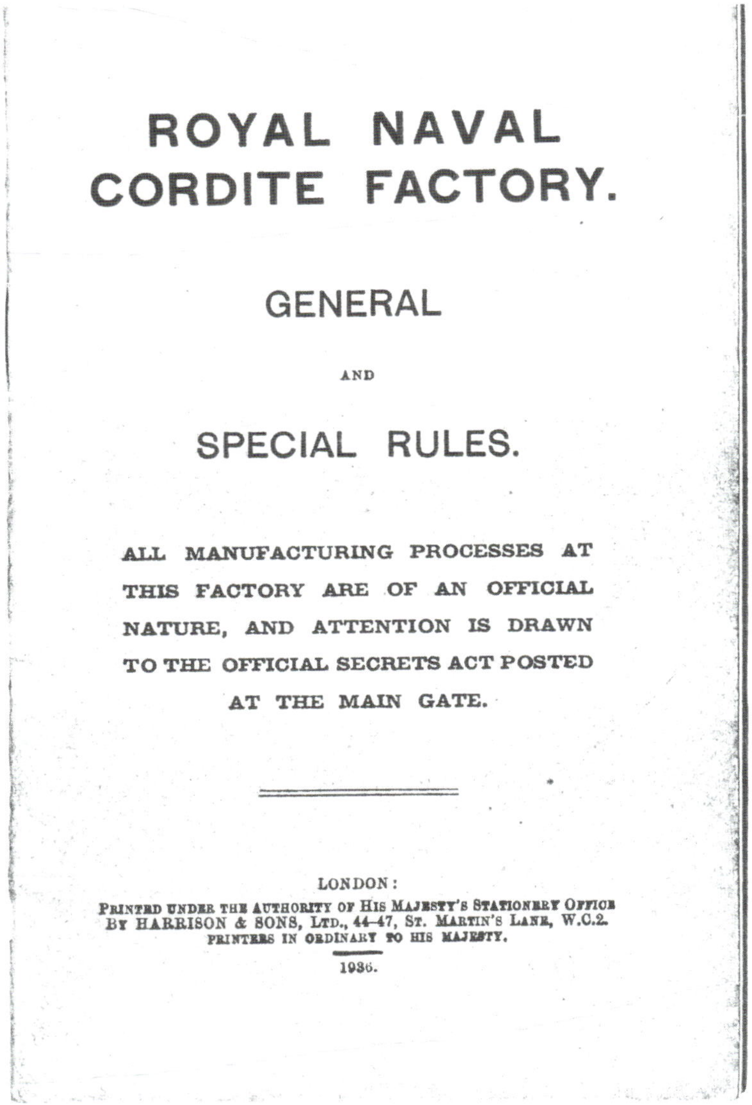

ROYAL NAVAL
CORDITE FACTORY.

GENERAL

AND

SPECIAL RULES.

ALL MANUFACTURING PROCESSES AT THIS FACTORY ARE OF AN OFFICIAL NATURE, AND ATTENTION IS DRAWN TO THE OFFICIAL SECRETS ACT POSTED AT THE MAIN GATE.

LONDON:
PRINTED UNDER THE AUTHORITY OF HIS MAJESTY'S STATIONERY OFFICE
BY HARRISON & SONS, LTD., 44-47, ST. MARTIN'S LANE, W.C.2.
PRINTERS IN ORDINARY TO HIS MAJESTY.
1936.

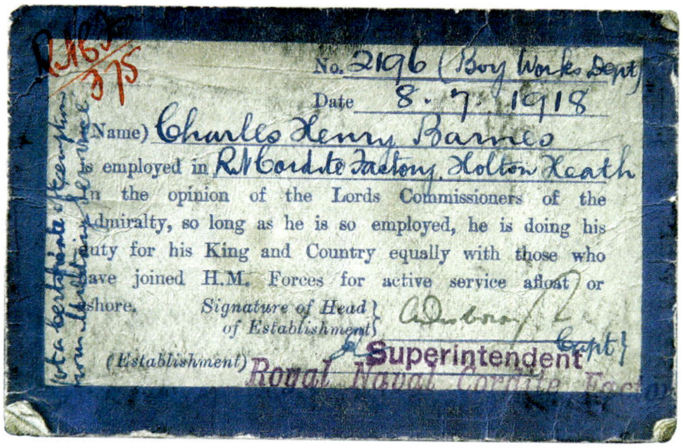

No. 2196 (Boy Works Dept)
Date 8. 7. 1918
(Name) Charles Henry Barnes
is employed in R.N. Cordite Factory, Holton Heath
In the opinion of the Lords Commissioners of the Admiralty, so long as he is so employed, he is doing his duty for his King and Country equally with those who have joined H.M. Forces for active service afloat or ashore.
Signature of Head of Establishment } A. Desborough Capt
Superintendent
(Establishment) Royal Naval Cordite Factory

R.N.C.F. Fire Drill Competition.

This is to Certify that

Mr. C. Barnes was a Member of the Workshops' Official Fire team which was successful in winning the 1924 Championship Cup for Fire Drill.

A. Desborough

Superintendent.

Above left: A First World War Certificate of Exemption from military service for a RNCF employee.
Above right: An inter-war fire-drill competition winners certificate.
Below left: A moral-boosting Second World War 'Salute The Soldier' savings scheme certificate.
Below Right: A rare survivor: a hand-written menu for the 1946 RNCF Home Guard annual dinner at the British Restaurant in Poole.

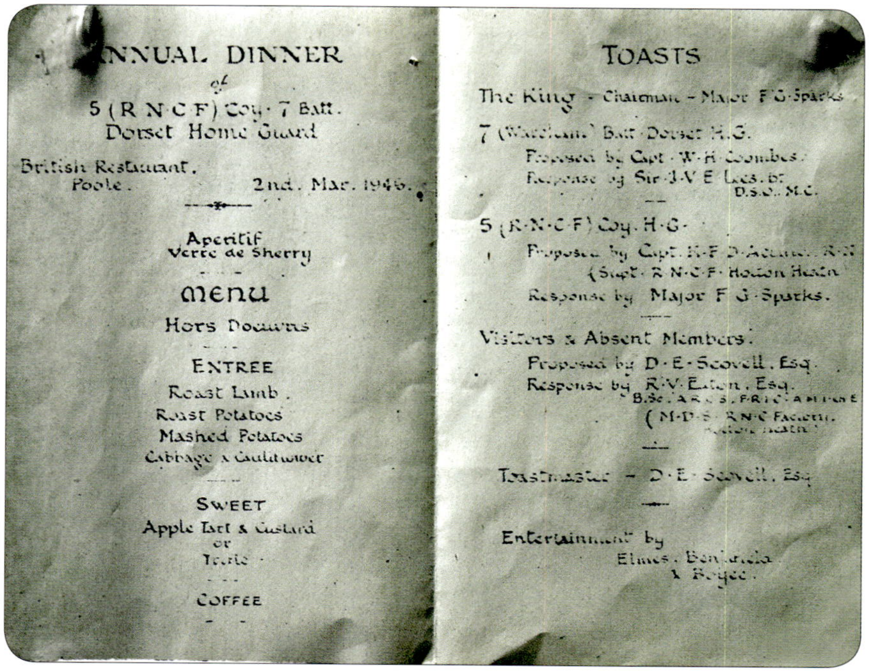

ANNUAL DINNER
of
5 (R N C F) Coy. 7 Batt.
Dorset Home Guard
British Restaurant,
Poole. 2nd. Mar. 1946.

Aperitif
Verre de Sherry

Menu
Hors Doeuvres

ENTREE
Roast Lamb.
Roast Potatoes
Mashed Potatoes
Cabbage & Cauliflower

SWEET
Apple Tart & Custard
or
Trifle

COFFEE

TOASTS
The King - Chairman - Major F.G. Sparks
7 (Wareham) Batt. Dorset H.G.
Proposed by Capt. W.H. Coombes.
Response by Sir J.V.E. Lees. Bt D.S.O., M.C.
5 (R.N.C.F.) Coy. H.G.
Proposed by Capt. H.F.D. Acland. R.N. (Supt. R.N.C.F. Holton Heath)
Response by Major F.G. Sparks.
Visitors & Absent Members.
Proposed by D.E. Scovell, Esq.
Response by R.V. Eaton, Esq. B.Sc. A.R.C.S. F.R.I.C. A.M.I.Chem.E (M.D.S. R.N.C. Factory Holton Heath)
Toastmaster - D.E. Scovell, Esq.
Entertainment by
Elmes, Bonfield & Boyce.

Above: Two typical, large and very colourful Second World War RNCF posters encouraging fund raising for the war effort. An idea of the size of the originals can be obtained from the picture of the RNCFA museum on page 213 where they are displayed upon the walls at the left of the picture.

APPENDIX

Detailed plans of Key sections

of the factory site

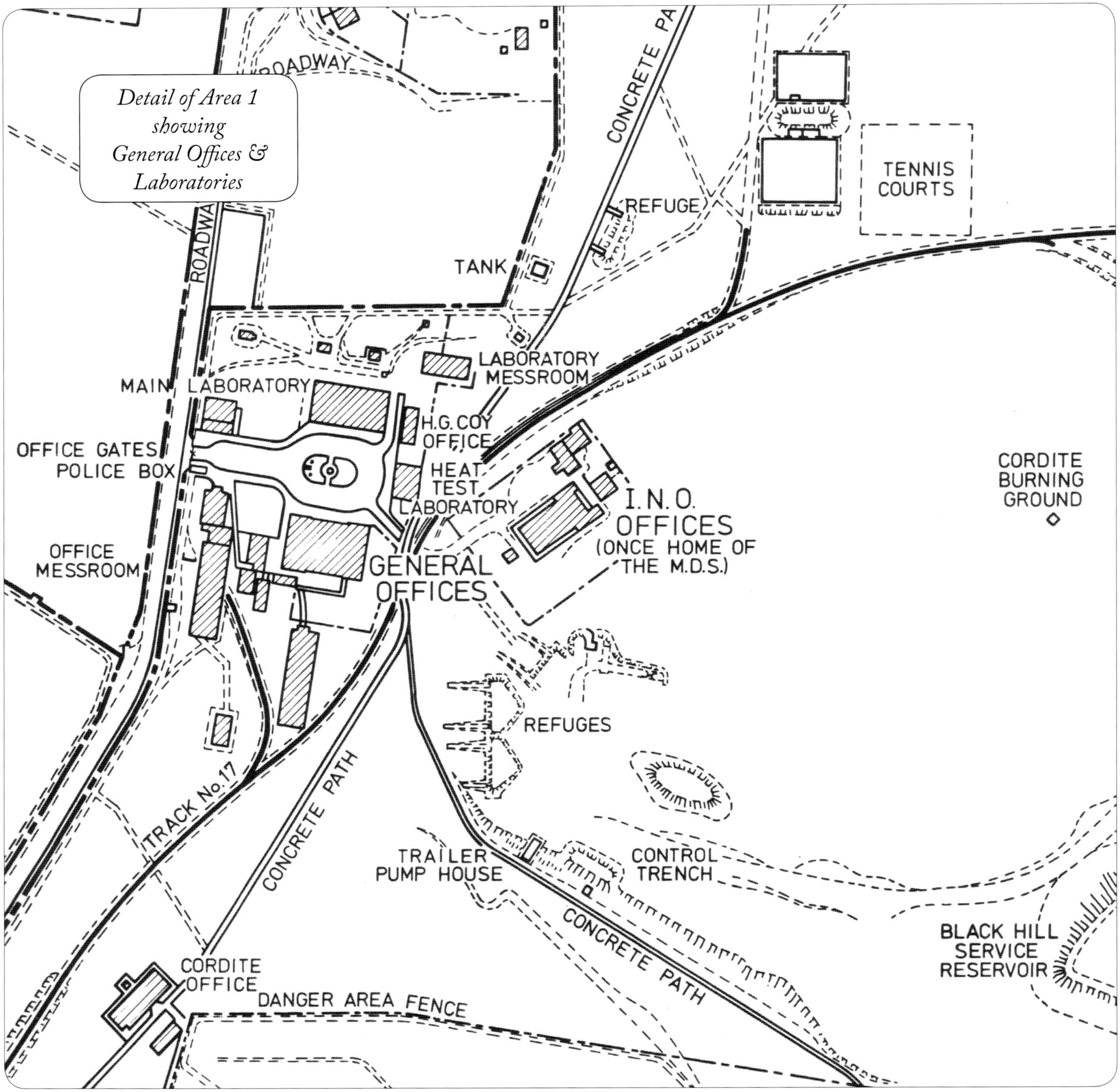
Detail of Area 1 showing General Offices & Laboratories
ROADWAY
ROADWAY
CONCRETE PATH
REFUGE
TENNIS COURTS
TANK
LABORATORY MESSROOM
MAIN LABORATORY
H.G. COY OFFICE
OFFICE GATES POLICE BOX
HEAT TEST LABORATORY
I.N.O. OFFICES (ONCE HOME OF THE M.D.S.)
CORDITE BURNING GROUND
OFFICE MESSROOM
GENERAL OFFICES
REFUGES
TRACK No.17
CONCRETE PATH
TRAILER PUMP HOUSE
CONTROL TRENCH
CONCRETE PATH
BLACK HILL SERVICE RESERVOIR
CORDITE OFFICE
DANGER AREA FENCE

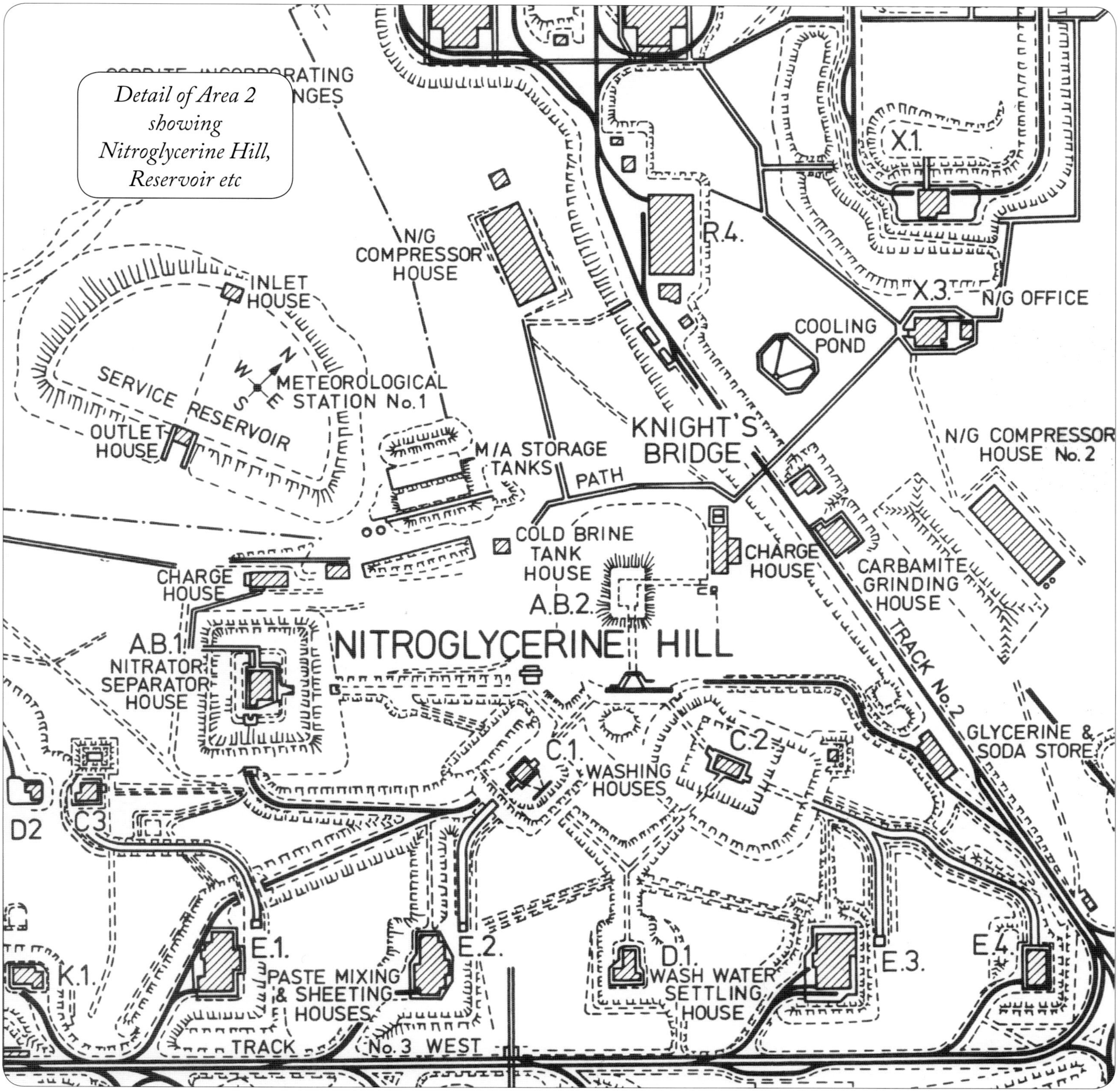

Detail of Area 2 showing Nitroglycerine Hill, Reservoir etc

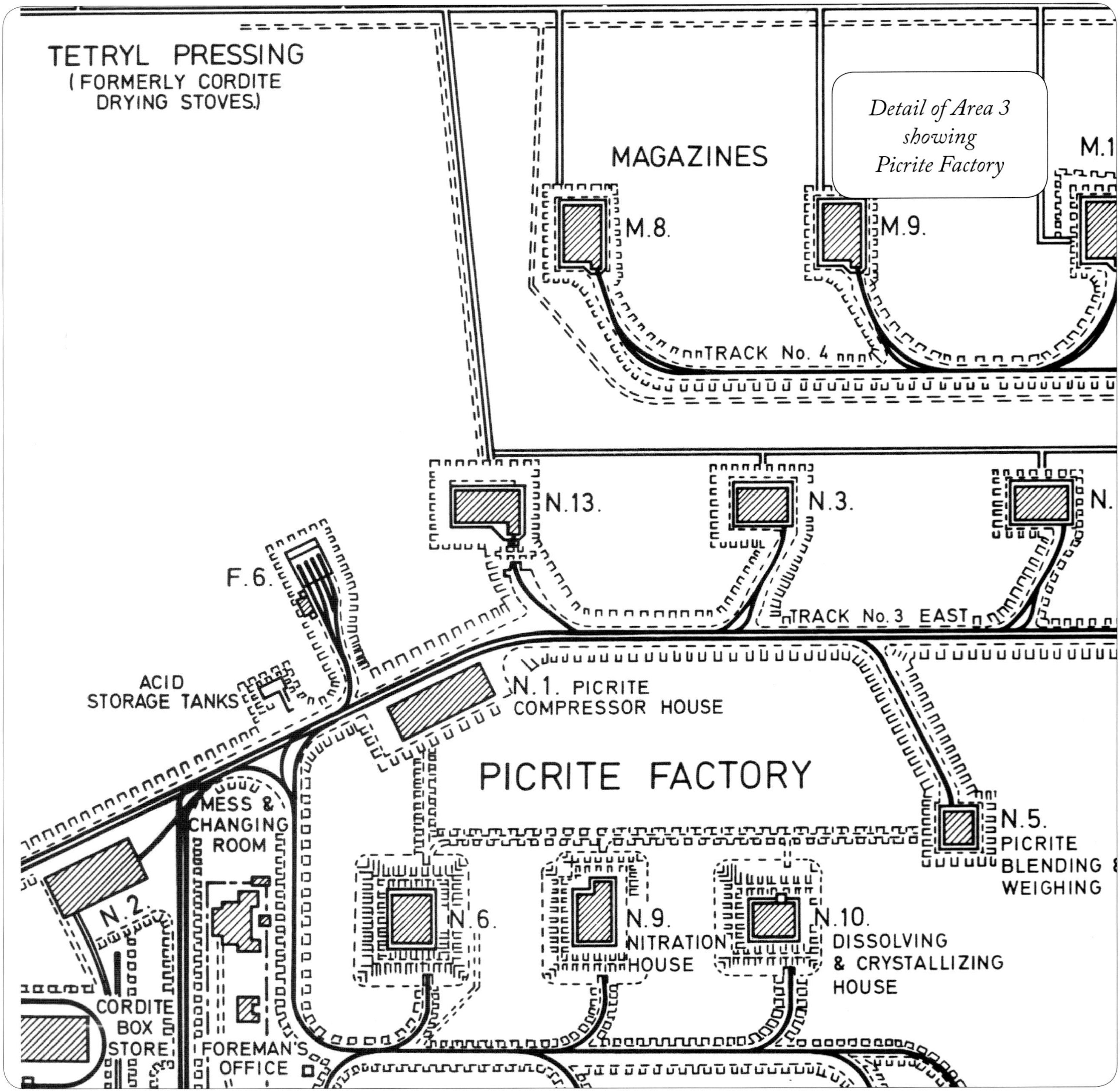
TETRYL PRESSING
(FORMERLY CORDITE
DRYING STOVES.)
MAGAZINES
Detail of Area 3
showing
Picrite Factory
M.8.
M.9.
TRACK No. 4
N.13.
N.3.
F.6.
TRACK No.3 EAST
ACID
STORAGE TANKS
N.1. PICRITE
COMPRESSOR HOUSE
PICRITE FACTORY
MESS &
CHANGING
ROOM
N.5.
PICRITE
BLENDING &
WEIGHING
N.2.
N.6.
N.9.
NITRATION
HOUSE
N.10.
DISSOLVING
& CRYSTALLIZING
HOUSE
CORDITE
BOX
STORE
FOREMAN'S
OFFICE

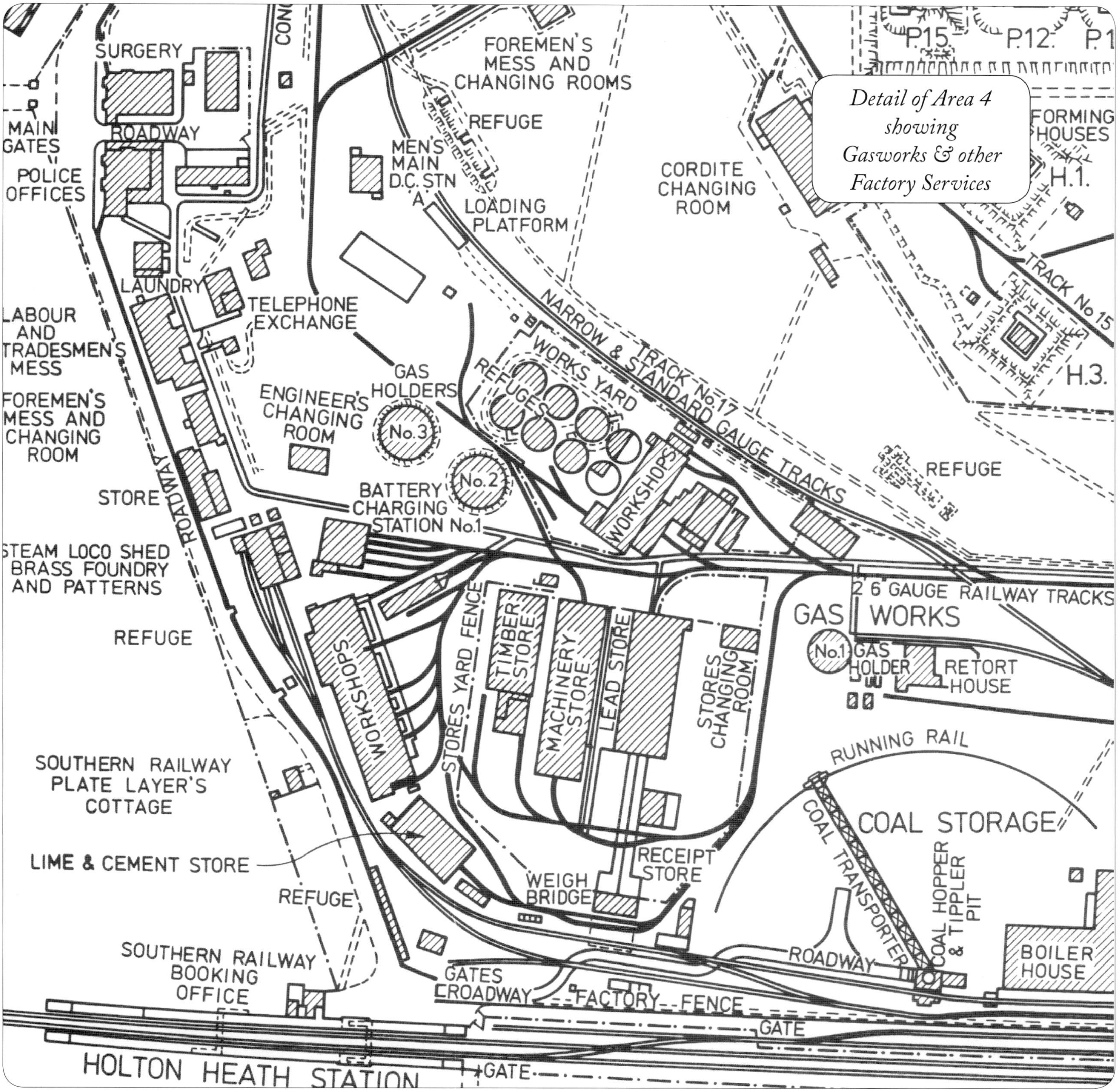

Detail of Area 4 showing Gasworks & other Factory Services

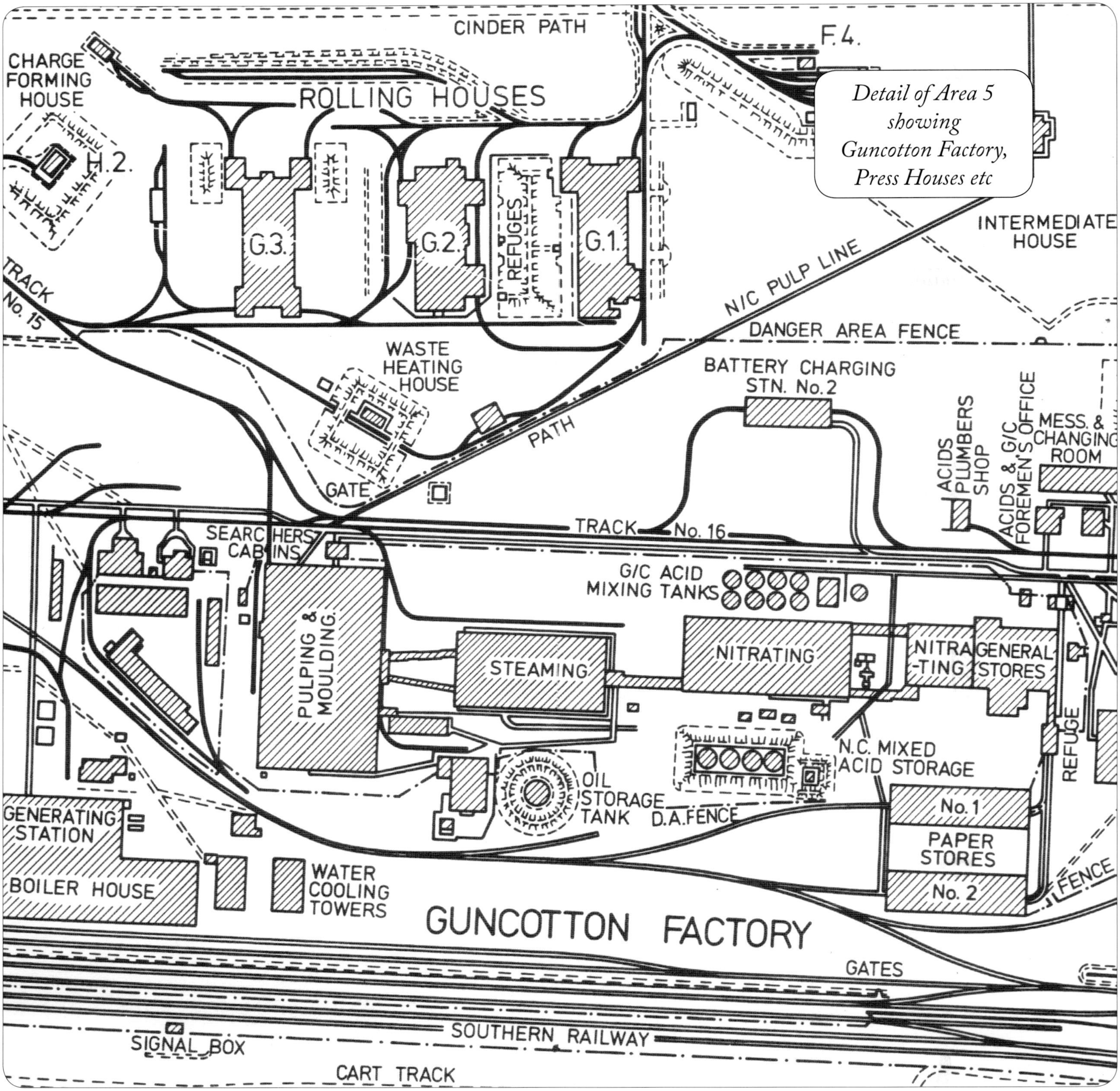

Detail of Area 5 showing Guncotton Factory, Press Houses etc

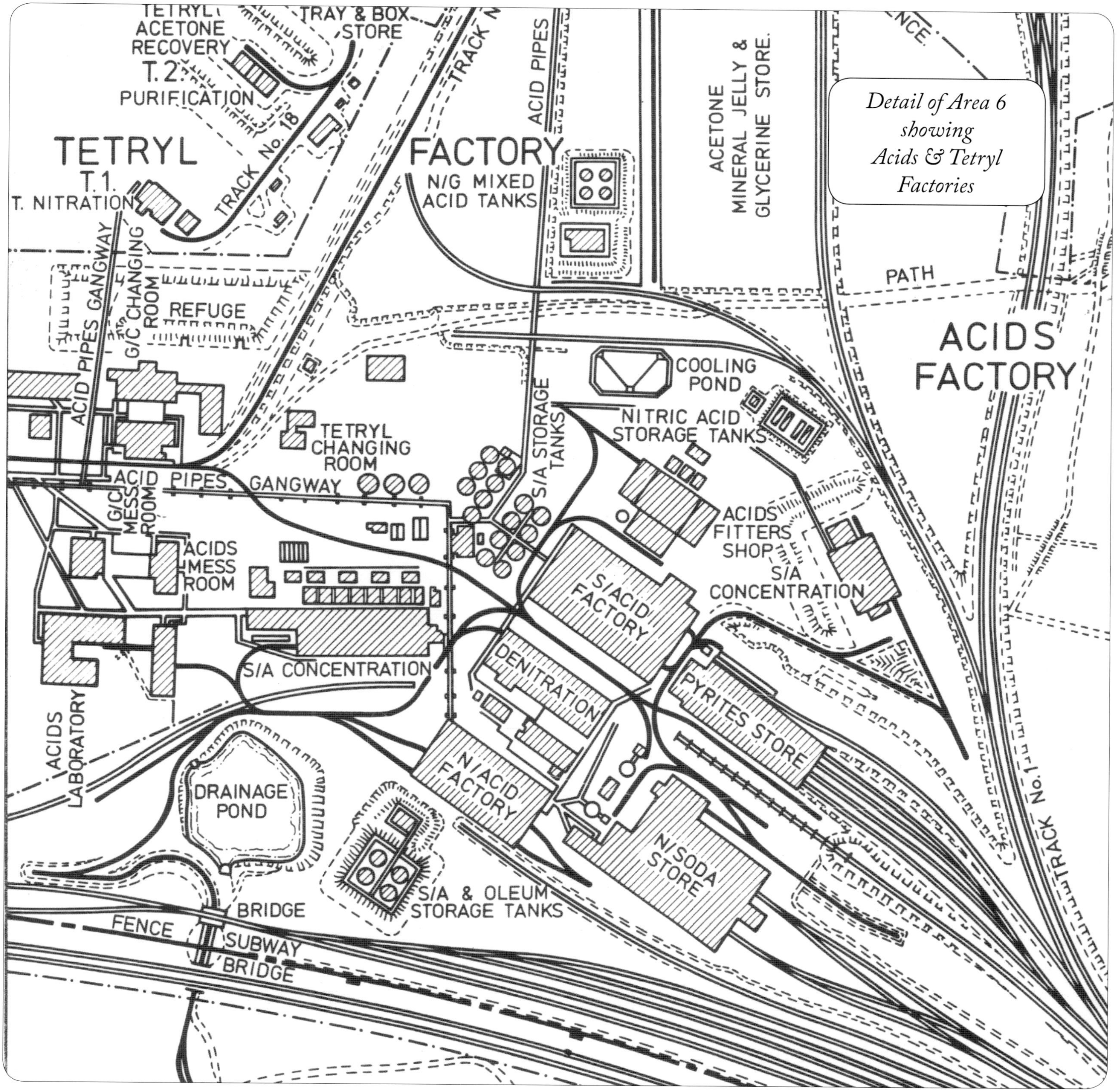

Detail of Area 6 showing Acids & Tetryl Factories

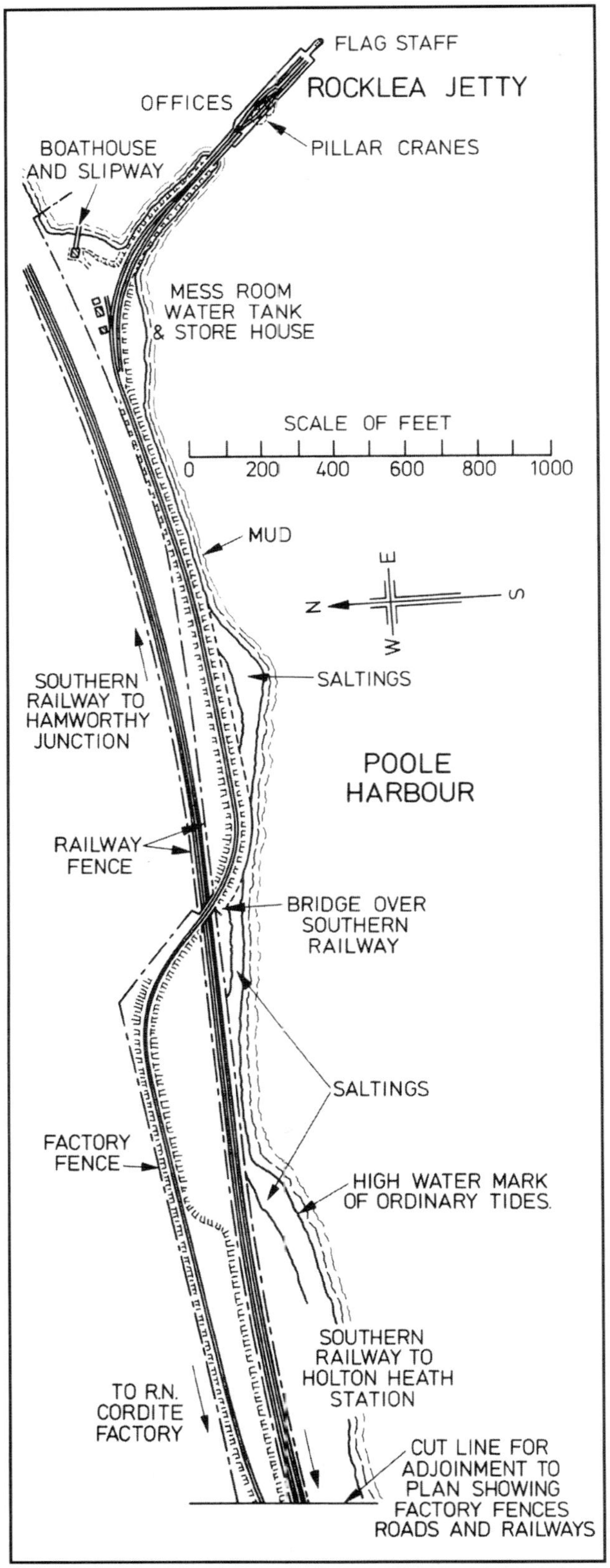
FLAG STAFF
ROCKLEA JETTY
OFFICES
PILLAR CRANES
BOATHOUSE
AND SLIPWAY
MESS ROOM
WATER TANK
& STORE HOUSE
SCALE OF FEET
0 200 400 600 800 1000
MUD
N
S
E
W
SALTINGS
SOUTHERN
RAILWAY TO
HAMWORTHY
JUNCTION
POOLE
HARBOUR
RAILWAY
FENCE
BRIDGE OVER
SOUTHERN
RAILWAY
SALTINGS
FACTORY
FENCE
HIGH WATER MARK
OF ORDINARY TIDES.
SOUTHERN
RAILWAY TO
HOLTON HEATH
STATION
TO R.N.
CORDITE
FACTORY
CUT LINE FOR
ADJOINMENT TO
PLAN SHOWING
FACTORY FENCES
ROADS AND RAILWAYS